D. Braun
Identification of Plastics

Dietrich Braun

Simple Methods for Identification of Plastics

With the Plastics Identification Table
by Hansjürgen Saechtling

Fourth Edition

Hanser Publisher, Munich
Hanser Gardner Publications, Inc., Cincinnati

Prof. Dr. rer. nat. Dietrich Braun
Deutsches Kunststoff-Institut, Schloßgartenstr. 6, D-64289 Darmstadt/Germany

Distributed in the USA and in Canada by
Hanser Gardner Publications, Inc.
6915 Valley Ave.
Cincinnati, OH 45244, USA
Fax: +1 (513) 527 8950
http://www.hansergardner.com

Distributed in all other countries by
Carl Hanser Verlag
Postfach 860420, D-81631 München, Germany
Fax: + 49 (89) 99 830-269
http://www.hanser.de

Library of Congress Cataloging-in-Publication Data
Braun, Dietrich, 1930 –
   [Erkennen von Kunststoffen. English]
   Simple methods for identification of plastics/Dietrich Braun;
with the plastics identification table by Hansjürgen Saechtling. – 4th ed.
   p.   cm.
   Includes bibliographical references and index.
   ISBN 1-56990-280-1
   1. Plastics – Analysis.   I. Title.
TP1140.B7213   1999
   668.4`197–dc21   99-15537
                                                  CIP

Die Deutsche Bibliothek – CIP-Einheitsaufnahme
Braun, Dietrich:
Simple methods for identification of plastics/Dietrich Braun.
With the plastics identification table/by Hansjürgen Saechtling. – 4. ed. –
Munich; Vienna: Hanser; Cincinnati: Hanser Gardner, 1999
   Dt. Ausg. u.d.T.: Braun, Dietrich: Erkennen von Kunststoffen
   ISBN 3-446-21113-6

© 1999 Carl Hanser Verlag Munich
Typeset in England by Alden Press, Oxford
Printed and bound in Germany by Wagner, Nördlingen

Preface

Processors and users of plastics often, for many different reasons, have to determine the chemical nature of a plastics sample. In contrast to plastics producers, however, they lack the specially equipped laboratories and the analytically trained staff for this purpose.

The complete identification of a high-molecular weight organic material is a rather complicated and often expensive problem. For many practical needs it is often sufficient to determine to which class of plastics an unknown sample belongs, for example to find out whether the material is a polyolefin or a polyamide. To answer such a question usually one only needs to use simple methods which do not require special chemical expertise.

In this book, now in its fourth edition, the author has compiled a selection of proven procedures which, based on his own experience, will enable the technician, the engineer and also the technical customer service representative to identify an unknown plastic, e.g. for purposes of quality control or plastics recycling. All described procedures were carried out by the author as well as by students in courses at the German Plastics Institute. Additional experience with these procedures was thus obtained and included in the book. The author welcomes any other comments and suggestions for additions by readers and users of this book.

Clearly one should not expect to obtain sophisticated information from these simple methods. In most cases one has to be satisfied with the identification of the plastic material, whereas the analysis of sometimes very small amounts of fillers, plasticizers, stabilizers or other additives is only possible through the use of more advanced physical and chemical methods. Similarly, it is not possible with simple methods to identify with certainty such combinations as copolymers and polymer blends. In such cases more sophisticated methods of analysis are required.

The new edition contains a number of additions, e.g. sections on thermoplastic elastomers, high-temperature resistant thermoplastics, and polymer blends. There is also a new section on the identification of metal-containing stabilizers in PVC. Finally, a short overview was added of more advanced analytical methods in which the most important methods for molecular and supermolecular characterization are included. In recent years, among instrumental analytical methods, infrared spectroscopy has made considerable advances and reasonably priced and user-friendly IR-spectrometers are now available. Therefore, the new edition includes the IR-spectra of the most important plastics.

The good reception that previous English, German, Spanish, and French editions of this book have received from users and from reviewers in various plastics journals and magazines, shows that inspite of all the modern analytical methods and advances in instrumental analysis there is still a need for simple methods for the identification of plastics. The analytical procedures described in this book do not require special chemical knowledge but they do require skills in carrying out simple operations. It is most important to remember to be careful in handling chemicals, solvents and open flames; other precautions will be pointed out in the pertinent sections. The necessary equipment is listed in the Appendix. With most experiments it is recommended that parallel experiments with known materials are carried out (a Plastics ID Kit is available through the Society of Plastics Engineers).

It is to be hoped that the present book fills the gap between the extensive plastics analysis volumes covering various methods in great detail and the tabular compilations of selected samples.

Naturally, this entails a compromise between investing in a greater experimental effort or being satisfied with the more limited information that can be obtained from simple qualitative analytical methods.

The development and trying out of the methods described in this book were part of the research programs at the German Plastics Institute and the author is grateful for the financial support of the Arbeitsgemeinschaft Industrieller Forschungsvereinigungen (AIF), Köhn and of the German Federal Ministry of Economics. The author also thanks the various persons who collaborated on these projects, especially Dr. R. Disselhoff, Dr. H. Pasch and Dipl.-Ing. E. Richter and also Ms. Ch. Hock who obtained the IR-spectra. As in previous editions, the author thanks the Carl Hanser Verlag and especially Dr. W. Glenz for their collaboration and taking the author's wishes into consideration.

Darmstadt   *Dietrich Braun*

Contents

1 Plastics and Their Characteristics

Plastics are high molecular weight (macromolecular or polymeric) organic substances that have usually been synthesized from low molecular weight compounds. They may also have been obtained by chemical modification of high molecular weight natural materials (especially cellulose). The raw materials are most often petroleum, natural gas, and coal. They can be reacted with air, water, or sodium chloride to prepare reactive monomers. The most important industrial synthetic processes for the preparation of plastics from monomers may be classified according to the mechanism of the formation reaction of the polymer, such as polymerization and condensation reactions. Since several chemically identical or similar plastics can be prepared in several different ways and from different raw materials, this classification has little meaning for the analysis of unknown plastics samples. On the other hand, in addition to chemical investigations, the appearance of a plastic as well as its behavior on heating yields useful information for its identification.

There are physical interactions between the individual macromolecules that constitute a plastic material, just as there are between the molecules of a low molecular weight compound. These physical interactions are responsible for cohesion and related properties such as strength, hardness, and softening behavior. Plastics that consist of linear threadlike molecules (several hundred nanometers* (nm) long and a few tenths of a

---

* $1\,\mathrm{nm} = 10^{-9}\,\mathrm{m} = 10\,\mathrm{\AA} = 10^{-6}\,\mathrm{mm}$

nanometer in diameter), i.e. of macromolecules, that are not strongly crosslinked can usually be softened on heating. In many cases they melt. Thus, when a polymeric material is heated above a certain temperature, the macromolecules which are more or less oriented with respect to each other at low temperatures can glide past each other to form a melt of relatively high viscosity. Depending on the degree of order of the macromolecule in the solid state, it is possible to distinguish between partly crystalline and (mostly disordered) amorphous plastics (see Figure 1). This degree of order also has an effect on the behavior of the plastic on heating and on its solubility.

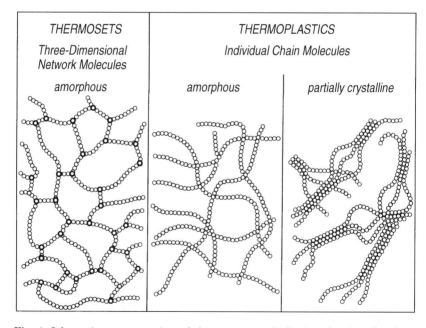

Fig. 1. Schematic representation of the structure of plastics, showing the three major types of macromolecular arrangements. Approximately 1 000 000 times actual size and greatly simplified. (Crystallites can also occur as the result of chain folding.)

Plastics that soften on heating and start to flow are called thermoplastics. On cooling, such plastics become solid again. This process can be repeated many times. There are several exceptions, as when the chemical stability (expressed in terms of the temperature at which chemical decomposition starts) is lower than the cohesion between the macromolecules due to interactions between the chains, in which case, on heating, the plastic undergoes chemical changes before it reaches its softening or melting point. A further indication, with few exceptions, that macromolecules are linear or branched is their solubility in many liquids, such as organic solvents. This process also reduces the interaction between the macromolecules; solvent molecules insert themselves between the polymer chains.

In contrast to thermoplastic materials, there are the so-called thermosetting materials. These, after processing into their final state, are crosslinked macromolecules that can neither melt nor dissolve. For such products one generally starts with liquid or soluble raw materials of a rather low molecular weight. These may be crosslinked by heating with or without pressure or through chemical reactions with additives and concurrent molding conditions. The results are crosslinked (hardened) high molecular weight materials in three-dimensional networks. These giant molecules can be broken down into smaller and therefore meltable and soluble fragments only by chemical destruction of the crosslinks. This may occur at rather high temperatures or with certain chemical reagents. Thermosets often contain fillers that may strongly influence the appearance and properties of the products.

Finally, from their physical appearance, we may distinguish the elastomers, rubberlike elastic materials consisting of usually relatively weakly crosslinked macromolecules. Crosslinkages of natural or synthetic rubber are formed during the molding or vulcanization process. Because of their crosslinked nature, elastomers do not melt on heating until just below their decomposition temperature. In this sense they behave differently from many other elastic thermoplastic materials such as plasticized polyvinyl chloride (PVC). In contrast to the chemically crosslinked elastomers (rubber), e.g. by sulfur or peroxides, network formation in the so-called thermoplastic elastomers (TPE) occurs through physical interactions between the macromolecules. On heating the physical interaction forces between the chain molecules decrease, so that these polymers can be thermoplastically shaped, and on cooling, as the physical interaction between the molecules becomes stronger, the material again behaves like an elastomer.

Table 1 lists the most important characteristics of these three groups of polymeric materials. In addition to elasticity, behavior on heating, density, and solubility can be used to differentiate between these materials. However, it should be kept in mind that fillers, pigments, or reinforcing agents, for example carbon black or glass fibers, lead to considerable deviations from these properties. Therefore it is not always possible to identify polymeric materials on the basis of these criteria. The densities listed in Table 1 are only rough approximations for some solid materials. For example, foams have densities of approximately $0.1\,\text{g/cm}^3$ or less. Structural foams with integral skin and cellular cores have densities between 0.2 and $0.9\,\text{g/cm}^3$ and often cannot be recognized as foams from their outer appearance.

It is not possible to discuss here the special properties of all the different types of plastic materials that can occur within these three groups. The plastics industry today, by employing copolymerization or chemical modification, is capable of producing an extraordinary number of combinations of properties, making the identification of corresponding plastics more complicated. Its physical appearance and its classification as a thermoplastic, thermoset, or elastomer therefore permit us to draw conclusions about the chemical nature of the plastic only in simple cases. But they often provide a useful additional way of characterizing the material.

In the last few years a number of products consisting of a mixture of different plastics have made their appearance; they are usually called polymer blends and polymer alloys. Their identification using simple methods presents considerable difficulties because

Table 1 Comparison of Different Classes of Plastics

	Structure	Physical Appearance*	Density (g/cm³)	Behavior on Heating	Behavior on Treating with Solvents
Thermo-plastics	Linear or branched macromolecules	Partially crystalline: flexible to horn-like; hazy, milky to opaque; only thin films are transparent	0.9–1.4 (except PTFE: 2–2.3)	Material softens; fusible and becomes clear on melting; often fibers can be drawn from the melt; heat-sealable (exceptions exist)	May swell; usually difficult to dissolve in cold solvents, but usually readily dissolved on heating the solvent, e.g., polyethylene in xylene
		Amorphous: colorless; clear and transparent without additives; hard to rubbery (e.g., on adding plasticizers)	0.9–1.9		Soluble (with few exceptions) in certain organic solvents, usually after initial swelling
Thermosets (after processing)	(Usually) tightly crosslinked macromolecules	Hard; usually contain fillers and are opaque. Without fillers they are transparent	1.2–1.4; filled: 1.4–2.0	Remain hard and almost dimension-ally stable until chemical decomposition sets in	Insoluble, do not swell or only slightly
Elastomers**	(Usually) lightly crosslinked macromolecules	Rubber-elastic and stretchable	0.8–1.3	Do not flow until close to temperature where chemical decomposition occurs	Insoluble, but will often swell

* A rough measure for the hardness of a plastic is its behavior on scratching with a fingernail: hard plastics scratch the nail; hornlike plastics have about the same hardness; flexible or rubbery plastics can be scratched or dented with a fingernail.

** The behavior of thermoplastic elastomers is described on page 14.

flame tests and pyrolysis tests are usually not unambiguous. Also a separation into different groups according to the pH-value of the pyrolysates does not permit a definite conclusion. In some cases, however, it is possible to separate polymer mixtures into their components if these have different solubility characteristics and then to identify the components (see Section 6.3).

For example, the examination of mixtures of polyamides and polyolefins is relatively easy because the polyamide component can be degraded by acid hydrolysis and the resulting low molecular weight fragments can then be identified according to the procedure described in Section 6.2.10. Table 3 lists some of the most important polymer blends together with their trade names and suppliers.

Although synthetic fibers and synthetic elastomers have the same chemical structure as plastics, they are not included among the latter group. Their identification will therefore be treated in this book only if they also occur as plastics. For example, polycaprolactam (Nylon 6) is used both for fiber production and as a molding material (see Sections 6.2.10 and 6.2.20).

Tables 2–5 contain a compilation of the plastics discussed in this volume, their chemical abbreviations, and some selected trade names. An extensive table of polymer acronyms on ASTM, DIN and ISO standards can be found in Appendix 8.5.

Table 2 Thermoplastics

Chemical or Technical Name	Abbreviation (Acronym)	Repeating Unit	Selected Trade Names (Registered Trademarks)
Polyolefins			
Polyethylene	PE	$-CH_2-CH_2-$	Alathon, Dowlex, Eraclene, Escorene, Finathene, Fortiflex, Fortilene, Hostalen, Lupolen, Marlex, Novapol, Moplen, Petrothene, Rexene, Riblene, Sclair, Tenite, Tuflin
Ethylene copolymers	EEA	With ethyl acrylate	Primacor, Lucalen
	EVA	With vinyl acetate	Elvax, Rexene, Ultrathene
Chlorinated polyethylene	PEC		Hostapren, Hypalon, Kelrinal
Chlorosulfonated polyethylene	CSM		Hypalon
Polypropylene	PP	$-CH_2-\underset{\underset{CH_3}{\mid}}{CH}-$	Adflex, Escorene, Fortilene, Moplen, Novolen, Petrothene, Pro-Fax, Rexene, Rexflex, Stamylan, Tenite, Valtec, Vestolen P
Polybutene-1	PB	$-CH_2-\underset{\underset{CH_2-CH_3}{\mid}}{CH}-$	Duraflex
Polyisobutylene	PIB	$-CH_2-\underset{\underset{CH_3}{\mid}}{\overset{\overset{CH_3}{\mid}}{C}}-$	Vistanex
Poly-4-methyl-pentene-1	PMP	$-CH_2-\underset{\underset{CH_2-\underset{\underset{CH_3}{\mid}}{CH}-CH_3}{\mid}}{CH}-$	(Mitsui Petrochem.) TPX
Styrene Polymers and Copolymers			
Polystyrene	PS	$-CH_2-CH-$ ⬡	Edistir, Ladene, Novacor, Styron

Table 2   Thermoplastics (continued)

Chemical or Technical Name	Abbreviation (Acronym)	Repeating Unit	Selected Trade Names (Registered Trademarks)
Modified polystyrene (high impact)	SB	Grafted with polybutadiene	Avantra, Edistir, Novacor, Styron
		Grafted with EPDM	
Styrene copolymers	SAN	With acrylonitrile	Luran, Lustran, Tyril
	ABS	Terpolymers: AN, B, S	Cycolac, Lustran, Magnum, Novodur, Polylac, Terluran, Toyolac
	ASA	Terpolymers: AN, S, acrylate	Centrex, Geloy, Luran
Halogen-Containing Polymers			
Polyvinyl chloride (rigid and flexible)	PVC	$-CH_2-\underset{\underset{Cl}{\mid}}{CH}-$	Benvic, Corvic, Dural, Geon, Unichem, Vestolit, Vinnolit
Modified PVC (high impact)	—	With EVA copolymers (EVA/VC graft copolymers) With chlorinated polyethylene With polyacrylate	
Polyvinylidene chloride	PVDC	$-CH_2-CCl_2-$	Ixan, Saran
Polytetrafluoroethylene	PTFE	$-CF_2-CF_2-$	Algoflon, Fluon, Halon, Hostaflon, Polyflon, Teflon
Polytetrafluoroethylene copolymers	ETFE	Copolymers with ethylene	Tefzel
	FEP	Copolymers with hexafluoropropylene	Neoflon, Teflon
Polytrifluorochloroethylene	CTFE	$-CF_2-\underset{\underset{Cl}{\mid}}{CF}-$	Kel-F, Fluorothene, Neoflon, Teflon
Trifluorochloroethylene copolymers	ECTFE	Copolymers with ethylene	Fluor, Halar

Table 2   Thermoplastics (continued)

Chemical or Technical Name	Abbreviation (Acronym)	Repeating Unit	Selected Trade Names (Registered Trademarks)
Halogen-Containing Polymers (continued)			
Perfluoroalkoxy-polymers	PFA	$-CF_2-CF_2-CF-CF_2-$ OR	Neoflon, Teflon
Polyvinyl fluoride	PVF	$-CH_2-CH-$ F	Kynar, Tedlar
Polyvinylidene fluoride	PVDF	$-CH_2-CF_2-$	Floraflon, Kynar, Solef
Polyacrylates and Polymethacrylates			
Polyacrylonitrile	PAN	$-CH_2-CH-$ CN	Barex
Polyacrylates	—	$-CH_2-CH-$ COOR R from different alcohols	
Polymethyl methacrylate	PMMA	CH_3 $-CH_2-C-$ $COOCH_3$	Acrylite, Degalan, Lucite, Perspex, Plexiglas
Methyl methacrylate copolymers	AMMA	Copolymers with AN	
Polymers with Heteroatom Chain Structure			
Polyoxymethylene	POM	$-CH_2-O-$	Celcon, Delrin, lupital, Tenac, Ultra-form
Polyphenylene oxide	PPO/PPE	CH_3 $-\bigcirc-O-$ CH_3	PPO
Modified PPO/PPE		With polystyrene or polyamide	Noryl, Luranyl, Prevex

Table 2 Thermoplastics (continued)

Chemical or Technical Name	Abbreviation (Acronym)	Repeating Unit	Selected Trade Names (Registered Trademarks)
Polymers with Heteroatom Chain Structure (continued)			
Polycarbonate	PC	$-\bigcirc-\overset{\overset{\displaystyle CH_3}{|}}{\underset{\underset{\displaystyle CH_3}{|}}{C}}-\bigcirc-O-CO-O-$	Apec, Calibre, Lexan, Makrolon
Polyethylene terephthalate	PET	$-CH_2-CH_2-O-CO-\bigcirc-CO-O-$	Dacron, Eastapak, Hiloy, Impet, Kodapak, Petlon, Petra, Rynite, Valox
Polybutylene terephthalate	PBT	$-(CH_2-CH_2)_2-O-CO-\bigcirc-CO-O-$	Celanex, Crastin, Pibiter, Pocan, Rynite, Ultradur, Valox
Polyamide	PA		
Polyamide 6 (Nylon 6)	PA 6	$-NH(CH_2)_5CO-$	Beetle, Capron, Celanese, Durethan, Grilon, Nypel, Ultramid
Polyamide 6,6 (Nylon 6,6)	PA 66	$-NH(CH_2)_6NH-CO(CH_2)_4CO-$	Capron, Technyl, Ultramid, Vydyne, Zytel
Polyamide 6,10 (Nylon 6,10)	PA 610	$-NH(CH_2)_6NH-CO(CH_2)_8CO-$	Amilan, Ultramid, Zytel
Polyamide 11 (Nylon 11)	PA 11	$-NH(CH_2)_{10}CO-$	Rilsan B
Polyamide 12 (Nylon 12)	PA 12	$-NH(CH_2)_{11}CO-$	Grilamid, Rilsan A, Vestamid
Aromatic PA		With terephthalic acid	Trogamid
Polyphenylene sulfide	PPS	$-\bigcirc-S-$	Fortron, Ryton, Suprec
Polysulfone	PSU	$-\bigcirc-\overset{\overset{\displaystyle O}{\|}}{\underset{\underset{\displaystyle O}{\|}}{S}}-\bigcirc-O-$	Radel, Ultrason
Polyethersulfone	PES	$-\bigcirc-\overset{\overset{\displaystyle CH_3}{|}}{\underset{\underset{\displaystyle CH_3}{|}}{C}}-\bigcirc-O-\bigcirc-\overset{\overset{\displaystyle O}{\|}}{\underset{\underset{\displaystyle O}{\|}}{S}}-\bigcirc-O-$	Udel

Table 2 Thermoplastics (continued)

Chemical or Technical Name	Abbreviation (Acronym)	Repeating Unit	Selected Trade Names (Registered Trademarks)
Cellulose Derivatives			

Cellulose (R=H)
– acetate (R=COCH$_3$)
– acetobutyrate
– propionate
  (R=CO–CH$_2$–CH$_3$)
– nitrate (R=NO$_2$)
Methyl cellulose
  (R=CH$_3$)
Ethyl cellulose
  (R=C$_2$H$_5$)
	CA
	CAB
	CP
	
	CN
	MC
	
	EC
			Tenite

Resins, Dispersions and Other Specialty Products			
Polyvinyl acetate	PVAC	$-CH_2-\underset{\underset{O-CO-CH_3}{\mid}}{CH}-$	Vinylite
Vinyl acetate copolymers		VAC/maleate VAC/acrylate VAC/ethylene	
Polyvinyl alcohol	PVAL	$-CH_2-\underset{\underset{OH}{\mid}}{CH}-$	Vinex
Polyvinyl ether		$-CH_2-\underset{\underset{OR}{\mid}}{CH}-$ R = different groups	Lutonal
Polyvinyl acetals	PVB	With butyraldehyde	Butvar, Butacite PVB
	PVFM	With formaldehyde	Formvar
Silicones	SI	$-\underset{\underset{R}{\mid}}{\overset{\overset{R}{\mid}}{Si}}-O-$ R = e.g., CH$_3$	Baysilone, Silastic (Resins, coating resins, oils, elastomers under different names, some may be hardened)
Casein	CS	–NH–CO– (polypeptide from milk albumin crosslinked with formaldehyde)	

Table 3   Polymer Blends (Selection)

Chemical or Technical Name	Abbreviation (Acronym)	Selected Trade Names (Registered Trademarks)
ABS-Blends		
with PA	ABS + PA	Triax
PC	ABS + PC	Bayblend, Cycoloy, Iupilon, Pulse, Terblend
PVC	ABS + PVC	Lustran ABS, Novaloy, Royalite
TPU	ABS + TPU	Desmopan, Estane, Prevail
ASA-Blends		
with PC	ASA + PC	Bayblend, Geloy, Terblend A
PVC	ASA + PVC	Geloy
PBT-Blends		
with ASA	PBT + ASA	Ultradur S
	PBT + PET	Valox
PET		
PC-Blends		
with PBT	PC + PBT	Azloy, Iupilon, Valox, Xenoy
PET	PC + PET	Makroblend, Sabre, Xenoy
PS-Blends		
with PE	PS + PE	Styroblend
PP	PS + PP	Hivalloy
PP-Blend		
with EP(D)M	PP + EP(D)M	Keltan, Santoprene
PPO-Blends		
with PS	PPO + PS	Noryl, Luranyl
PA	PPO + PA	Noryl GTX

Table 4 Thermosets

Chemical or Technical Designation	Abbreviation (Acronym)	Starting Materials	Reactive Groups* or Curing Agent	Intermediate Products and Curing Procedures
Phenoplasts				
Phenolic resins	PF	Phenol (R=H) and substituted phenols (e.g., cresols and formaldehyde)	$-CH_2OH$ OH ⬡–R	Novolacs (not self-curing; cured, e.g., by addition of hexamethylenetetramine) Resoles (cured under pressure and heating, sometimes with catalysts, to Resits)
Cresol-formaldehyde resins	CF	Cresol (R=CH₃) and formaldehyde		
Aminoplasts				
Urea-formaldehyde resins	UF	Urea (sometimes also thiourea)	$-NH_2$ $-NH-CH_2OH$ $-N(CH_2OH)_2$	Intermediate products in the form of aqueous solutions or solids; curing occurs under pressure and heating, sometimes using acid catalysts
Melamine-formaldehyde resins	MF	Melamine and formaldehyde		
Unsaturated polyester resins	UP	Polyesters with unsaturated dicarboxylic acids, usually maleic acid, and saturated acids such as succinic acid, adipic acid, phthalic acid, and diols such as butanediol	$-CO-CH=CH-CO-$	Polyesters are usually dissolved in styrene, seldom in other monomers; curing by radical copolymerization using hot- or cold-type initiators
Glass fiber-reinforced unsaturated polyester resins	GUP or GF-UP			

* Since the chemical composition of crosslinked plastics cannot be given with any accuracy, this table lists starting materials and reactive groups but does not describe the products or give the trade names of the many available thermosets that differ in composition as well as content of additives, e.g., fillers.

Table 4 Thermosets (continued)

Chemical or Technical Designation	Abbre-viation (Acro-nym)	Starting Materials	Reactive Groups* or Curing Agent	Intermediate Products and Curing Procedures (continued)
Epoxy resins	EP	From di- or polyols or bis-phenols and epichlorohydrin or other epoxide-forming components	$-CH-CH-$ with O forming epoxide ring	Liquid or solid intermediates that are cured either hot, e.g., using dicarboxylic acids or anhydrides, or cold, using, e.g., di- or polyamines
Polyurethanes	PUR	Di- or polyisocyanates react with diols or polyols to form crosslinked hard or soft (usually elastic) products	$-N{=}C{=}O + HO- \rightarrow$ $NH-CO-O-$	Isocyanates (e.g., MDI, TDI, Desmodur) and OH-containing compounds (different polyols) are reacted in the liquid or molten state

* Since the chemical composition of crosslinked plastics cannot be given with any accuracy, this table lists starting materials and reactive groups but does not describe the products or give the trade names of the many available thermosets that differ in composition as well as content of additives, e.g., fillers.

Table 5 Elastomers*

Chemical or Technical Elastomers	Abbreviation (Acronym)	Starting Materials	Typical Repeating Units	
Polybutadiene	BR	Butadiene	$-CH_2-CH=CH-CH_2-$	1,4-Addition (*cis* or *trans*) 1,2-Addition (isotactic, syndiotactic, or atactic)
			$-CH_2-CH-$ $CH=CH_2$	
Polychloroprene (Neoprene, Perbunan)	CR	Chloroprene	$-CH_2-C=CH-CH_2-$ Cl	Structural isomers exist
Polyisoprene	PIP NR	Isoprene Natural rubber	$-CH_2-C=CH-CH_2-$ CH_3	*cis*-1,4-Polyisoprene (Guttapercha or balata: *trans*-1,4-Polyisoprene)
Nitrile rubber	NBR	Acrylonitrile and butadiene		
Styrene-butadiene rubber	SBR	Styrene and butadiene		
Butyl rubber	IIR	Isobutylene and a small amount of isoprene		
Ethylene-propylene rubber	EPM EPDM or EPD	Ethylene and propylene Terpolymers with dienes		

* This table contains only a selection of the most important elastomers. Their structure is shown in the unvulcanized state.

Table 5   Elastomers* (continued)

Chemical or Technical Elastomers	Abbreviation (Acronym)	Starting Materials	Typical Repeating Units
Fluorine rubber	FE	Fluorine-containing olefins	
Chlorohydrin rubber	CHR	Epichlorohydrin-ethylene oxide copolymers	
Propylene oxide rubber	POR	Copolymer from propylene oxide and allyl glycidyl ether	

* This table contains only a selection of the most important elastomers. Their structure is shown in the unvulcanized state.

2 General Introduction to the Analysis of Plastics

2.1 Analytical Procedure

Each plastic analysis begins with screening tests. In addition to the observation of several characteristics such as solubility, density, softening, and melting behavior, an important role is played by heating in a combustion tube (pyrolysis test) and in an open flame (flame test). If these preliminary tests do not yield a positive identification, examine the materials for the presence of hetero-atoms such as nitrogen, halogens (especially chlorine and fluorine), and sulfur. Then begin a systematic analysis by testing the solubility, and proceed to simple specific tests. In addition, try to identify the possible presence of organic or inorganic fillers or other additives such as plasticizers or stabilizers. Unfortunately, the simple approaches discussed here seldom give reliable information about the type and amounts of such additives.

As an aid in the identification of the type of plastic used in semifinished plastic materials or in plastic moldings, the "Plastics Identification Table" by Dr. Hj. Saechtling included in this book (Section 8.1) has proved to be quite useful. Starting from the appearance of the material and its elastic behavior, the table leads to a series of simple tests which allow further differentiation between types of plastics. Procedures used in these tests, mentioned in the headings of the table, are described in detail at appropriate places in the text of this book. For such tests it is sufficient to take small splinters or filings removed from the sample at some inconspicuous place.

2.2 Sample Preparation

Plastics as a raw material usually are in the form of powders, granules, and very occasionally dispersions. After processing, they are usually encountered as films, plates, profiles, or molded products.

Some screening tests, for example, a flame test, can be carried out on the original form (granules, chips, etc.). For most tests, however, it is better if the sample is available in a finely divided or powdery state. To reduce the size of the particles, use a mill; a coffee grinder may be sufficient. On thorough chilling by adding dry ice (solid CO_2), most tough or elastic materials become brittle and can be ground. The chilling prevents them from becoming overheated during the grinding process.

Very often, processed plastic materials contain additives: plasticizers, stabilizers, fillers, or coloring agents such as pigments. Such additives usually do not interfere with the simple, not very specific, screening tests. For a quantitative determination or for the definite identification of a plastic material, the additives must first be removed. For this purpose, extraction (see Figure 2) or precipitation methods are used. Processing aids such as stabilizers or lubricants similar to plasticizers can usually be extracted with ether or other organic solvents. If an extraction apparatus (Soxhlet) is not available, it may be sufficient to shake the finely divided sample with ether or to heat the sample in ether for several hours under reflux. Use extreme caution. Ether is flammable. Do not use an open flame.

Linear polymers can be separated from fillers or reinforcing agents (glass fibers or carbon black) by dissolving them in suitable solvents. (For the selection of solvents, see Section 3.1.)

Fig. 2. Soxhlet extractor. The extraction liquid is heated to boiling in a round-bottom flask, and the resulting vapor is condensed in a reflux condenser mounted at the top of the extractor. From the condenser the liquid drops onto the solid sample in the cup. When the liquid in the extractor vessel reaches the exit tube near the top (right side of the extractor) it flows back into the round-bottom flask. The solvent must have a lower specific gravity than the material being extracted (otherwise the sample would float out of the extractor cup).

All insoluble material then remains behind and can be isolated by filtration. The dissolved polymer can be reprecipitated by adding the solution dropwise to a 5–10 times larger volume of the precipitating agent. As a precipitating agent, methanol is nearly always suitable. In some cases water can be the precipitating agent.

Crosslinked plastics cannot be separated from fillers in this way due to their insolubility. Inorganic fillers (glass fibers or chalk) can sometimes be isolated by burning the sample in a porcelain cup, although this is not always the case. Carbon black may also burn off. However, it is frequently necessary to try special methods, which vary from case to case.

3 Screening Tests

3.1 Solubility

Among the many solvents for plastics, the most widely used are toluene, tetrahydrofuran, dimethylformamide, diethyl ether, acetone, and formic acid. In certain cases, chloroethylene, ethyl acetate, ethanol, and water are also useful. It should be noted that the flammability and toxicity of many solvents requires special care in handling. Benzene should be avoided as far as possible. Tables 6 and 7 show a compilation of the behavior of the most important plastics in various solvents. For the systematic analysis of plastics, the distinction between soluble and insoluble polymers provides a first separation into two groups. We can then apply chemical methods to investigate these two groups further.

For the determination of solubility, add approximately 0.1 g of the finely divided plastic to a test tube with 5–10 ml of the solvent. Over the course of several hours, thoroughly shake the test tube and observe the possible swelling of the sample. This can often take quite a long while. If necessary, heat the test tube gently with constant agitation. This can be done with a Bunsen burner, but a water bath is better. Great care must be employed to avoid sudden boiling up of the solvent and having it spray out of the test tube, since most organic solvents or their vapors are flammable. If the solubility test leaves doubts

Table 6   Solubility of Plastics

Polymer	Solvent	Nonsolvent
Polyethylene, polybutene-1, isotactic polypropylene	p-Xylene*, trichlorobenzene*, decane*, decalin*	Acetone, diethyl ether, lower alcohols
Atactic polypropylene	Hydrocarbons, isoamyl acetate	Ethyl acetate, propanol
Polyisobutylene	Hexane, toluene, carbon tetrachloride, tetrahydrofuran	Acetone, methanol, methyl acetate
Polybutadiene, polyisoprene	Aliphatic and aromatic hydrocarbons	Acetone, diethyl ether, lower alcohols
Polystyrene	Toluene, chloroform, cyclohexanone, butyl acetate, carbon disulfide	Lower alcohols, diethyl ether (swells)
Polyvinyl chloride	Tetrahydrofuran, cyclohexanone, methyl ethyl ketone, dimethyl formamide	Methanol, acetone, heptane
Polyvinyl fluoride	Cyclohexanone, dimethyl formamide	Aliphatic hydro-carbons, methanol
Polytetrafluoroethylene	Insoluble	—
Polyvinyl acetate	Chloroform, methanol, acetone, butyl acetate	Diethyl ether, petroleum ether, butanol
Polyvinyl isobutyl ether	Isopropanol, methyl ethyl ketone, chloroform, aromatic hydrocarbons	Methanol, acetone
Polyacrylates and polymethacrylates	Chloroform, acetone, ethyl acetate, tetrahydrofuran, toluene	Methanol, diethyl ether, petroleum ether

---

* Often soluble only at elevated temperatures.

Table 6   Solubility of Plastics (continued)

Polymer	Solvent	Nonsolvent
Polyacrylamide	Water	Methanol, acetone
Polyacrylic acid	Water, dilute alkalies, methanol, dioxane, dimethylformamide	Hydrocarbons, methanol, acetone, diethyl ether
Polyvinyl alcohol	Water, dimethylformamide*, dimethylsulfoxide*	Hydrocarbons, methanol, acetone, diethyl ether
Cellulose	Aqueous cupriammonium hydroxide, aqueous zinc chloride, aqueous calcium thiocyanate	Methanol, acetone
Cellulose diacetate	Acetone	Methylene chloride
Cellulose triacetate	Methylene chloride, chloroform, dioxane	Methanol, diethyl ether
Methyl cellulose (trimethyl)	Chloroform	Ethanol, diethyl ether, petroleum ether
Carboxymethyl cellulose	Water	Methanol
Aliphatic polyesters	Chloroform, formic acid	Methanol, diethyl ether, aliphatic hydrocarbons
Polyethylene terephthalate	m-Cresol, o-chlorophenol, nitrobenzene, trichloroacetic acid	Methanol, acetone, aliphatic hydrocarbons
Polyacrylonitrile	Dimethylformamide, dimethylsulfoxide, concentrated sulfuric acid	Alcohols, diethyl ether, water, hydrocarbons

* Often soluble only at elevated temperatures.

Table 6   Solubility of Plastics (continued)

Polymer	Solvent	Nonsolvent
Polyamides	Formic acid, conc. sulfuric acid, dimethylformamide, m-cresol	Methanol, diethyl ether, hydrocarbons
Polyurethanes (uncrosslinked)	Formic acid, γ-butyrolactone, dimethylformamide, m-cresol	Methanol, diethyl ether, hydrocarbons
Polyoxymethylene	γ-Butyrolactone*, dimethylformamide*, benzyl alcohol*	Methanol, diethyl ether, aliphatic hydrocarbons
Polyethylene oxide	Water, dimethylformamide	Aliphatic hydrocarbons, diethyl ether
Polydimethylsiloxane	Chloroform, heptane, diethyl ether	Methanol, ethanol

* Often soluble only at elevated temperatures.

and/or if insoluble particles (glass fibers or inorganic fillers) remain behind, they must be removed. They can be most easily filtered or decanted after the solution has been allowed to stand overnight. For the test, evaporating a part of the supernatant liquid on a watchglass leaves the dissolved material as a residue. The filtered solution can also be dropped into a nonsolvent for that particular plastic, in which case dissolved polymer will precipitate. Petroleum ether or methanol and occasionally water are used as precipitating agents.

The solubility of a plastic material depends very much on its chemical structure and to some extent on the size of the molecules (molecular weight). The solvents mentioned in Table 7, therefore, do not always permit an unambiguous identification.

Table 7   Plastics Dissolved by Selected Solvents

Water	Tetrahydrofuran (THF)	Boiling Xylene	Dimethyl-formamide (DMF)	Formic Acid	Insoluble in all of these solvents
Polyacrylamide Polyvinyl alcohol Polyvinyl methyl ether Polyethylene oxide Polyvinyl-pyrrolidone Polyacrylic acid	All uncrosslinked polymers*	Polyolefins Styrene polymers Vinyl chloride polymers Polyacrylates Polytrifluoro-chloroethylene	Polyacrylonitrile Polyformalde-hyde (in boiling DMF)	Polyamides Polyvinyl alcohol derivatives Urea- and melamine-formaldehyde condensates (uncured)	Polyfluoro hydrocarbons Polyethylene terephthalate** Crosslinked (cured, vulcanized) polymers

* Except polyolefins, polyfluoro hydrocarbons, polyacrylamide, polyoxymethylene, polyamides, polyethylene terephthalate, polyurethanes, urea and melamine resins.

** Soluble in nitrobenzene.

3.2 Density

The density ϱ, is the quotient of mass M and volume V of a material

$$\varrho = \frac{M}{V} \, g/cm^3$$

With plastics the density is seldom useful as a means of characterization. Many processed plastics contain hollow spaces, pores, or imperfections. In such cases (for example, foams), the quotient of the mass and the volume determined by the outer boundaries of the sample is determined as a raw density according to ASTM D792. The true density can in principle be determined from the mass and the true volume.

With compact solids it is often sufficient to measure a single sample for the determination of the volume. With plastics in powder or granular form, the volume is determined by measuring the amount of displaced liquid in a pycnometer or by means of buoyancy measurements. In all cases one needs relatively accurate weighings, especially with small amounts of sample.

For many purposes it is simpler to use the flotation procedure in which the sample is made to float in a liquid of the same density. The density of the liquid may then be determined according to known methods with an aerometer. One can use aqueous zinc chloride or magnesium chloride solutions as the liquids. With densities below $1 \, g/cm^3$, methanol-water mixtures are useful.

For density to be determined according to the flotation process, of course, the sample must not dissolve or swell in the liquid, and it must wet completely. Make sure that no air bubbles appear on the surface of the sample for they may affect the measurement. Any bubbles must be completely removed. Carbon black, glass fibers, and other fillers can also influence the density measurements greatly. For example, the densities can vary depending on the filler content from $0.98 \, g/cm^3$ (polypropylene with 10% by weight of talcum) to $1.71 \, g/cm^3$ (polybutyleneterephthalate containing

Table 8   Approximate Densities of Important Plastics

Density (g/cm^3)	Material
0.80	Silicone rubber (silica filled up to 1.25)
0.83	Polymethylpentene
0.85–0.92	Polypropylene
0.89–0.93	High-pressure (low-density) polyethylene
0.91–0.92	Polybutene-1
0.91–0.93	Polyisobutylene
0.92–1.0	Natural rubber
0.94–0.98	Low-pressure (high-density) polyethylene
1.01–1.04	Nylon 12
1.03–1.05	Nylon 11
1.04–1.06	Acrylonitrile-butadiene-styrene copolymers (ABS)
1.04–1.08	Polystyrene
1.05–1.07	Polyphenylene oxide
1.06–1.10	Styrene-acrylonitrile copolymers
1.07–1.09	Nylon 610
1.12–1.15	Nylon 6
1.13–1.16	Nylon 66
1.1–1.4	Epoxy resins, unsaturated polyester resins
1.14–1.17	Polyacrylonitrile
1.15–1.25	Cellulose acetobutyrate
1.16–1.20	Polymethyl methacrylate
1.17–1.20	Polyvinyl acetate
1.18–1.24	Cellulose propionate
1.19–1.35	Plasticized PVC (approx. 40% plasticizer)
1.20–1.22	Polycarbonate (based on bisphenol A)
1.20–1.26	Crosslinked polyurethanes
1.24	Polysulfone
1.26–1.28	Phenol-formaldehyde resins (unfilled)
1.21–1.31	Polyvinyl alcohol
1.25–1.35	Cellulose acetate
1.30–1.41	Phenol-formaldehyde resins filled with organic materials (paper, fabric)
1.3–1.4	Polyvinyl fluoride
1.34–1.40	Cellulose nitrate
1.38–1.41	Polyethylene terephthalate
1.38–1.41	Rigid PVC
1.41–1.43	Polyoxymethylene (polyformaldehyde)
1.47–1.52	Urea- and melamine-formaldehyde resins with organic fillers
1.47–1.55	Chlorinated PVC
1.5–2.0	Phenoplasts and aminoplasts with inorganic fillers
1.7–1.8	Polyvinylidene fluoride
1.8–2.3	Polyester and epoxy resins filled with glass fibers
1.86–1.88	Polyvinylidene chloride
2.1–2.2	Polytrifluoromonochloroethylene
2.1–2.3	Polytetrafluoroethylene

50% glass fibers). Foams cannot be characterized by density determinations.

If more accurate methods for the determination of the density are not available, immerse the sample in methanol (density, ϱ, at $20\,^\circ\text{C} = 0.79\,\text{g/cm}^3$), water ($\varrho = 1\,\text{g/cm}^3$), saturated aqueous magnesium chloride solution ($\varrho = 1.34\,\text{g/cm}^3$), or saturated aqueous zinc chloride solution ($\varrho = 2.01\,\text{g/cm}^3$). Then observe whether the sample stays on the surface of the liquid, floats inside it, or sinks. Its behavior indicates whether it has a lower or a higher density than the liquid in which it is immersed. Table 8 contains the densities of the most important plastics (of course, there can be variations).

To prepare a saturated solution, add chemically pure zinc chloride or magnesium chloride in small portions and with shaking or stirring to distilled water until, on further addition, the material dose not dissolve and a residue remains at the bottom. The solution process is relatively slow, and the saturated solutions are rather viscous.

For the preparation of a 1-liter saturated solution one needs approximately 1575 g zinc chloride or 475 g magnesium chloride. Both solutions are hygroscopic and therefore must be kept in closed flasks.

3.3 Behavior on Heating

Linear or branched, that is, not crosslinked, thermoplastic materials usually first begin to soften on heating and then on further heating (amorphous polymers) begin to flow over a rather ill-defined temperature range (see Figure 3). Partially crystalline plastics in general have narrow melting ranges, which, however, are usually less sharply defined than the melting points of low molecular weight crystalline materials. Above the flow temperature the sample begins to break down chemically (pyrolysis). This process of thermal degradation produces low molecular weight fragments, which are often flammable or have

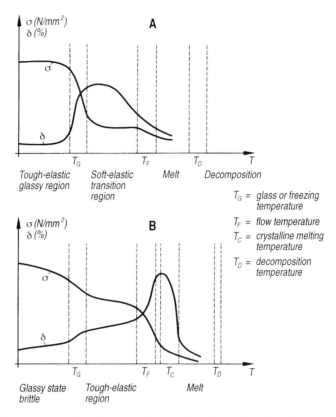

Fig. 3. Dependence of tensile strength σ and of elongation δ on temperature for amorphous thermoplastics (A) and partially crystalline thermoplastics (B).

a characteristic odor. Thermosets and elastomers show little or no flow up to their decomposition temperature (see Figure 4). At that point they also form many typical degradation products, which give important information for the identification of plastic.

In addition to pyrolysis, flame tests yield useful information, since the behavior in the flame shows characteristic differences depending on the nature of the plastic. Pyrolysis tests and flame tests are therefore among the most important screening tests in the analysis of a plastic. They often permit direct conclusions so that one can then begin with specific tests.

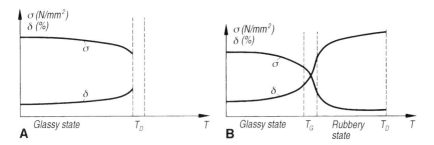

Fig. 4. Dependence of tensile strength σ and elongation δ on temperature for thermosets (A) and for elastomers (B).

3.3.1 Pyrolysis Tests

To examine the behavior of a plastic on heating without direct exposure to flame, add a small sample (approx. 100 mg) to a pyrolysis tube, gripping the upper end of the tube with a clip or pair of tongs. At the open end of the tube place a piece of moist litmus or pH paper. In some cases one inserts a wad of loose cotton wool or glass wool that has been moistened with water or methanol in the open end of the pyrolysis tube. Heat the test tube in the flame of a Bunsen burner that has been reduced to minimum heat, taking care to point the open end of the tube away from the face. (*Caution*: Wear safety glasses.) Heating should occur so slowly that the changes in the sample and in the smell of the decomposition gases can be properly determined.

Depending on the reaction of the escaping vapors with litmus it is possible to distinguish three different groups: acidic (the litmus paper turns red), neutral (no color change), or basic (the litmus paper turns blue). The pH paper is somewhat more sensitive. Table 9 shows the reactions of the decomposition products of the most important plastics. Depending on their composition some plastics can appear in the pyrolysis test in different groups, for example, phenolformaldehyde resins or polyurethanes.

Table 9   Litmus and pH Tests for Vapors of Plastics*

	Litmus Paper	
Red	Essentially unchanged	Blue

	pH Paper	
0.5–4.0	5.0–5.5	8.0–9.5
Halogen-containing polymers (e.g. PVC)	Polyolefins	Polyamides
Polyvinyl esters	Polyvinyl alcohol	ABS polymers
Cellulose esters	Polyvinyl acetals	Polyacrylonitrile
Polyethylene terephthalate	Polyvinyl ethers	Phenolic and cresol resins
Novolacs	Styrene polymers (included styrene-acrylonitrile copolymers)**	Amino resins (aniline-, melamine-, and urea-formaldehyde resins
Polyurethane elastomers	Polymethacrylates	
Unsaturated polyester resins	Polyoxymethylene	
Fluorine-containing polymers	Polycarbonates	
Vulcanized fiber	Linear polyurethanes	
Polyalkylene sulfide	Silicones	
	Phenolic resins	
	Epoxy resins	
	Crosslinked polyurethanes	

* Slowly heated in a pyrolysis tube.
** Some samples show slightly alkaline behavior.

3.3.2 Flame Tests

To test the behavior of the plastic in a flame, hold a small sample of the plastic with a pair of tweezers or a spatula in a low flame. Reduce the gas supply to the Bunsen burner to its minimum. Observe the flammability of the plastic in and out of the flame. Also note the formation of drops of burning or melting plastic as well as the odor after the flame is extinguished. The surface under the burner should be covered with aluminum foil to catch any falling droplets. Table 10 shows the behavior of the most important plastics in the flame test. However, the flammability of plastics is influenced strongly by the addition of flame-retarding additives, and therefore in practice results may deviate from these shown in Table 10.

Table 10 Behavior of Plastics on Burning (Flame Test)*

Flammability	Appearance of Flame	Odor of Vapors	Material
Does not burn	—		Silicones
	—	Stings (hydrofluoric acid, HF)	Polytetrafluoroethylene
	—		Polytrifluorochloroethylene
			Polyimides
Difficult to ignite, extinguishes when removed from flame	Bright, sooty	Phenol, formaldehyde	Phenolic resins
	Bright yellow	Ammonia, amines, formaldehyde	Amino resins
	Green edge	Hydrochloric acid	Chlorinated rubber
			Polyvinyl chloride
			Polyvinylidene chloride
			(without flammable plasticizers)
Burns in the flame, extinguishes slowly or not at all outside the flame	Shiny, sooty	—	Polycarbonates
	Yellow, grey smoke	—	Silicone rubber
	Yellow-orange, blue smoke	Burnt horn	Polyamides
	Yellow	Phenol, burnt paper	Phenolic resin laminates
	Shiny, material decomposes	Irritating, scratches the throat	Polyvinyl alcohol
	Yellow-orange	Burnt rubber	Polychloroprene
	Yellow-orange, sooty	Sweetly aromatic	Polyethylene terephthalate
	Yellow, blue edge	Stinging (isocyanate)	Polyurethanes
	Yellow, blue center	Paraffin	Polyethylene, polypropylene
	Shiny, sooty	Sharp	Polyester resins (glass fiber reinforced)
	Yellow	Phenol	Epoxy resins (glass fiber reinforced)

* For the behavior of high temperature-resistant thermoplastics see Section 6.2.19.

Table 10 Behavior of Plastics on Burning (Flame Test)* (continued)

Flammability	Appearance of Flame	Odor of Vapors	Material
Ignites readily, continues burning after flame is removed	Shiny, sooty	Sweetish, natural gas	Polystyrene
	Dark yellow, slightly sooty	Acetic acid	Polyvinyl acetate
	Dark yellow, sooty	Burnt rubber	Rubber
	Shiny, blue center, crackles	Sweetish, fruity	Polymethyl methacrylate
	Bluish	Formaldehyde	Polyoxymethylene
	Dark yellow, slightly soft	Acetic acid and butyric acid	Cellulose acetobutyrate
	Light green, sparks	Acetic acid	Cellulose acetate
	Yellow-orange	Burnt paper	Cellulose
	Bright, violent	Nitrogen oxides	Cellulose nitrate

* For the behavior of high temperature-resistant thermoplastics see Section 6.2.19.

For a systematic evaluation of flammability and odor tests, the scheme described by G. H. Domsch (Kunststoffe 61 (1971), p. 669) is recommended (see Figure 5).

3.3.3 Melting Behavior

As was mentioned previously, softening or melting occurs only with linear plastics. In some cases, however, the softening or melting range lies above the range in which polymers are thermally stable. In that case, decomposition starts before melting of the sample can be observed. With crosslinked plastics there is usually no softening until just below the point where chemical degradation occurs. Therefore, this kind of behavior is an indication, although not an unequivocal one, that the material is a cured thermoset (see Figure 4). In general, high molecular weight compounds do not have as sharp a melting point as crystalline low molecular weight organic compounds (see Figure 3).

The glass or freezing temperatures of polymers are very characteristic of specific polymers. These are the temperatures at which certain molecular segments become mobile, without entire chains being able to glide past each other so that a viscous flow can begin. The determination of the glass temperature is hardly possible with simple means, because for many plastics it lies considerably below room temperature. Among suitable methods are: differential thermel analysis, the measurement of temperature dependence of the refractive index, and mechanical properties such as the modulus of elasticity.

The softening range of a plastic can be determined with the usual methods of organic chemistry, either in a melting point tube or with a hot-stage microscope. A hot stage (Kofler stage) with which melting points can be determined to an accuracy of 2–3 °C (see Figure 6) is very useful. However, the resulting values often depend to a considerable extent on the rate of heating and on the presence of certain additives, especially plasticizers. The most reliable melting points are those of partially crystalline polymers.

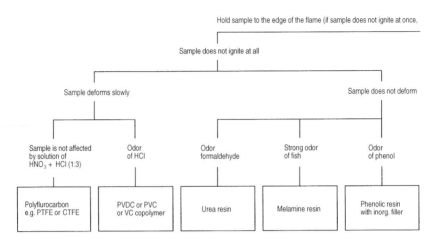

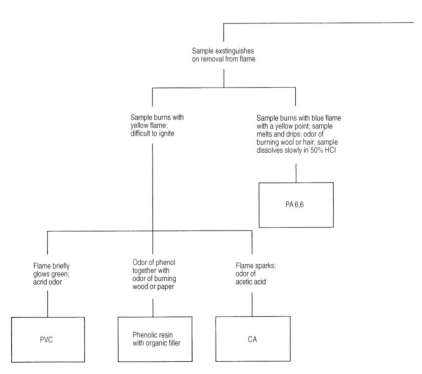

Fig. 5. Flammability and Odor Tests (after G. H. Domsch)

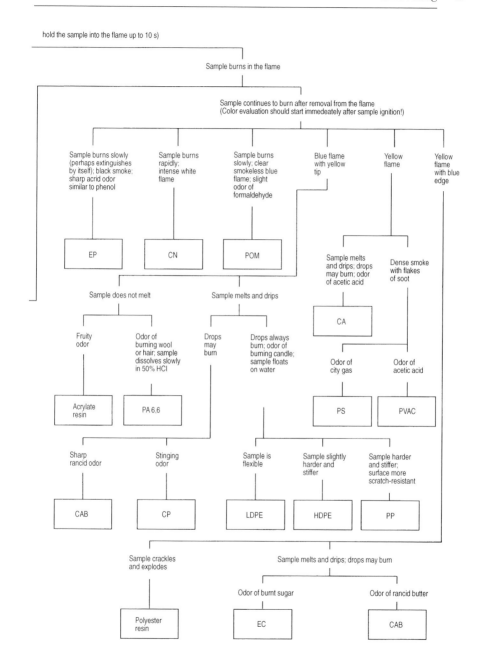

hold the sample into the flame up to 10 s)

Sample burns in the flame

Sample continues to burn after removal from the flame
(Color evaluation should start immedeately after sample ignition!)

Sample burns slowly (perhaps extinguishes by itself); black smoke; sharp acrid odor similar to phenol

Sample burns rapidly; intense white flame

Sample burns slowly; clear smokeless blue flame; slight odor of formaldehyde

Blue flame with yellow tip

Yellow flame

Yellow flame with blue edge

EP

CN

POM

Sample melts and drips; drops may burn; odor of acetic acid

Dense smoke with flakes of soot

Sample does not melt

Sample melts and drips

CA

Fruity odor

Odor of burning wool or hair; sample dissolves slowly in 50% HCl

Drops may burn

Drops always burn; odor of burning candle; sample floats on water

Odor of city gas

Odor of acetic acid

Acrylate resin

PA 6.6

PS

PVAC

Sharp rancid odor

Stinging odor

Sample is flexible

Sample slightly harder and stiffer

Sample harder and stiffer; surface more scratch-resistant

CAB

CP

LDPE

HDPE

PP

Sample crackles and explodes

Sample melts and drips; drops may burn

Odor of burnt sugar

Odor of rancid butter

Polyester resin

EC

CAB

The different polyamides, for example, can be easily distinguished (compare Section 6.2.10). Values for the most important plastics are shown in Table 11. A more comprehensive tabulation can be found in *Plastics Analysis Guide* by A. Krause, A. Lange, and M. Ezrin (Hanser Publishers, 1983).

As was previously mentioned, it is possible with the aid of thermoanalytical methods to obtain more accurate information about the thermal behavior of plastics. Foremost among these methods are thermogravimetry (TG) and differential scanning calorimetry (DSC). TG provides information about changes in

Table 11   Softening and Melting Ranges of Important Thermoplastics

Thermoplastic	Softening or Melting Range ($^\circ$C)
Polyvinyl acetate	35–85
Polystyrene	70–115
Polyvinyl chloride	75–90 (softens)
Polyethylene, density 0.92 g/cm^3	about 110
density 0.94 g/cm^3	about 120
density 0.96 g/cm^3	about 135
Polybutene-1	125–135
Polyvinylidene chloride	115–140 (softens)
Polymethyl methacrylate	120–160
Cellulose acetate	125–175
Polyacrylonitrile	130–150 (softens)
Polyoxymethylene	165–185
Polypropylene	160–170
Nylon 12	170–180
Nylon 11	180–190
Polytrifluorochloroethylene	200–220
Nylon 6, 10	210–220
Nylon 6	215–225
Polybutylene terephthalate	220–230
Polycarbonate	220–230
Polyethersulfone	228–230
Poly-4-methylpentene-1	230–240
Nylon 6, 6	250–260
Polyethylene terephthalate	250–260
Polyphenylene sulfide	260–280
Polyarylether ketone	340–380

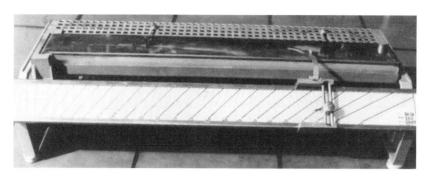

Fig. 6. Hot stage. A linear temperature gradient from 50 to 250 °C is created by resistance heaters along the metal bar. The sample (as finely divided as possible) is placed directly on the metal bar. The temperature at the border between solid powder and molten material can be determined directly from a scale on the hot stage.

weight on heating of a sample, for example through loss of moisture or evaporation of volatile products resulting from chemical degradation reactions of the macromolecules. In DSC one determines the specific heat of a plastic as a function of temperature by heating the sample in a programmed manner and determining the changes in energy content, i.e. enthalpy, of the sample. From the resulting curves it is possible to determine glass transition temperatures and melting temperatures and the corresponding specific heat changes. For these methods one needs rather expensive instruments and experienced personnel and for additional details the reader should consult the pertinent literature.

4 Testing for Heteroatoms

The previously described simple screening tests are not always sufficient to identify an unknown plastic with certainty. In some cases the use of chemical reactions for identification purposes cannot be avoided. First one tests for heteroatoms, those elements which are present in the plastic in addition to carbon and hydrogen, such as nitrogen, sulfur, chlorine, fluorine, silicon, and sometimes phosphorus. Unfortunately, there is no simple direct method for the certain identification of oxygen, so it is not possible to test for oxygen in a qualitative manner. The following reactions presuppose a certain amount of experimental skill and the necessary care.

For the qualitative determination of the elements nitrogen, sulfur, and chlorine, the *Lassaigne method* is usually used. Combine approximately 50–100 mg of a finely divided sample with a pea-sized piece of sodium or potassium in a pyrolysis tube. Heat this carefully in a Bunsen flame until the metal melts. (*Caution!* Wear safety glasses and keep the opening of the tube directed away from the eyes.) The sample must be free of water, which would react explosively with the metal. Sodium and potassium must be stored in oil or immersed in a similar inert hydrocarbon. When used, a small piece of the metal is held with tweezers and the required amount is cut off with a knife or a spatula on a piece of filter paper. Then carefully blot it dry with the filter paper. Use it immediately and return the remainder to the oil-containing bottle. In no case should the remainder be destroyed by throwing it into water.

After heating, carefully place the red-glowing tube in a small beaker with approximately 10 m distilled water. The glass tube will shatter and the reaction products will dissolve in the water. Unreacted metal will react with the water; therefore, carefully stir with a glass rod until no further reaction occurs. Then filter the nearly colorless liquid or remove the liquid by careful pipetting from the glass splinters and carbonized residues. For the following tests, use approximately 1–2 ml of this original solution.

- *Nitrogen*

Add a small amount (a spatula tip) of ferrous sulfate to a 1–2 ml sample of the original solution and boil it quickly. Let it cool and add a few drops of 1.5 % ferric chloride solution. After acidification with dilute hydrochloric acid, a precipitate of Berlin blue occurs. The presence of a small amount of nitrogen results in a light green solution from which a precipitate results only after standing for several hours. If the solution remains yellow, there is no nitrogen present.

- *Sulfur*

The original solution is reacted with an approximately 1 % aqueous sodium nitroferricyanide solution. A deep violet color indicates sulfur. This reaction is very sensitive; to confirm, add a drop of the alkaline solution of the sample under investigation to a silver coin. If sulfur is present, a brown spot of silver sulfide will form. An alternative procedure is to acidify the original solution with acetic acid (test with litmus or pH paper) and then add several drops of aqueous 2 M lead acetate solution or test with lead acetate paper. A back precipitate of lead sulfide or darkening of the paper indicates sulfur.

The presence of sulfur in polysulfides, polysulfones and in sulfur-vulcanized rubber can be demonstrated by the following somewhat uncertain test. The sample is heated in dry air (pyrolysis) and the gases formed during this process are bubbled through a dilute barium chloride solution. The presence of sulfur is indicated by a white precipitate of barium sulfate.

- *Chlorine*

This is a general test for heavier halogens, but bromine and iodine almost never occur in plastics. Acidify a sample of the original solution with dilute nitric acid and add a small amount of silver nitrate solution (2 g in 100 ml distilled water; keep the solution in the dark or in a brown flask). A white flaky precipitate that dissolves again on the addition of an excess of ammonia indicates the presence of chlorine. A light yellow precipitate that is difficult to dissolve in ammonia indicates the presence of bromine. A yellow precipitate that is insoluble in ammonia is characteristic for iodine.

• *Fluorine*

Acidify the original solution with dilute hydrochloric acid or acetic acid and then add a 1 N calcium chloride solution. A gel-like precipitate of calcium fluoride indicates the presence of fluorine (compare also below).

• *Phosphorus*

On the addition of a solution of ammonium molybdate to a portion of the original solution that was acidified with nitric acid, one obtains a precipitate on heating for approximately 1 min. To prepare the molybdate solution, dissolve 30 g ammonium molybdate in approximately 60 ml hot water, cool, and add water to make 100 ml. Then add a thin stream of a solution of 10 g ammonium sulfate in 100 ml 55 % nitric acid (from 16 ml water and 84 ml concentrated nitric acid). Let it stand for 24 hr, remove the supernatant by suction or by decantation, and keep the solution well sealed in the dark.

• *Silicon*

Mix approximately 30–50 mg of the plastic sample with 100 mg dry sodium carbonate and 10 mg sodium peroxide in a small platinum or nickel crucible (carefully). Melt it slowly over a flame. After cooling, dissolve the material in a few drops of water, bring it quickly to a boil, and neutralize or slightly acidify it with dilute nitric acid. Add 1 drop of molybdate solution (see phosphorus test), then heat nearly to boiling. Cool the sample, add 1 drop of benzidine solution (50 mg benzidine dissolved in 10 ml 50 % acetic acid; add water to make 100 ml), and then add 1 drop of saturated aqueous sodium acetate solution. A blue color indicates silicon.

Other Identification Reactions

Halogens, especially chlorine and bromine, can be easily identified with the very sensitive *Beilstein test*. Heat the end of a copper wire in a Bunsen flame until the flame is colorless. After cooling, put a small amount of the substance to be examined on the wire and heat it at the edge of the colorless part of the flame. When the plastic burns, the presence of halogen can be inferred if the flame is colored green or blue-green.

Fluorine can be demonstrated by placing approximately 0.5 g of the plastic in a small test tube and pyrolyzing it in a Bunsen flame. After cooling, add a few milliliters of concentrated sulfuric acid. The presence of fluorine is indicated by a characteristic non-wettability of the wall of the test tube. (Make a comparison experiment with a sample of known fluorine content.)

From the results of the tests for heteroatoms, useful conclusions can be drawn:

- *Chlorine*

occurs in plastics such as PVC, chlorinated polyethylene, and rubber hydrochloride. Some plasticizers also contain chlorine. Flameproofing agents often contain chlorine or bromine.

- *Nitrogen*

is found in polyamides, aminoplastics, cellulose nitrate, and in cellophane films treated with nitrogen-containing lacquers.

- *Sulfur*

when found in rubber elastic materials, indicates vulcanized rubber, polysulfones, or polysulfides.

- *Phosphorus*

is seldom found in plastics (with the exception of casein). However, it indicates the presence of phosphate plasticizers, stabilizers, or flame-proofing agents.

A compilation of the most important plastics containing hetero-atoms is shown in Table 12.

Table 12   Classification* of Plastics by their Heteroatoms

—	O			Heteroatoms					
	Cannot be saponified	Can be saponified**		Halogens	N, O	S, O	Si	N, S	N, S, P
		SN < 200	SN > 200						
Polyolefins	Polyvinyl alcohol	Natural resins	Polyvinyl acetate and copolymers	Polyvinyl chloride	Polyamides	Polyalkylene sulfide	Silicones	Thiourea condensates	Casein resins
Polystyrene	Polyvinyl ethers	Modified phenolic resins	Polyacrylates and polymethacrylates	Polyvinylidene chloride and copolymers	Polyurethanes Polyureas	Vulcanized rubber	Polysiloxanes	Sulfamide condensates	
Polyisoprene	Polyvinyl acetals		Polyesters	Polyfluoro-hydrocarbons	Aminoplastics				
Butyl rubber	Polyglycols		Alkyd resins	Chlorinated rubber	Polyacrylonitrile and copolymers				
	Polyaldehydes		Cellulose esters						
	Phenolic resins			Rubber hydrochloride	Polyvinylcarbazole				
	Xylene resins				Polyvinylpyrrolidone				
	Cellulose								
	Cellulose ethers								

* After W. Kupfer, Z. analyt. Chem. **192** (1963) 219.

** SN = Saponification number (amount of potassium hydroxide used in mg KOH per g of substance).

5 Analytical Procedures

On the basis of the screening tests described in previous chapters and with the use of certain specific reactions, the most important plastics can be identified through simple separation procedures. Test first for heteroatoms (Chapter 4), then for solubility in different solvents (Section 3.1). If necessary, test for other characteristic physical properties or carry out chemical reactions.

As has previously been pointed out, the solubility of plastics depends in many cases on the molecular weight. With copolymers and polymer mixtures it also depends on the composition, and this can lead to problems. In that case, it is necessary to use additional, more complicated tests.

Plastics can be classified into four groups according to the elements present. Group I contains chlorine or fluorine, group II contains nitrogen, group III contains sulfur, and group IV contains no identifiable heteroatoms.

For the following solubility determination (procedure given in Section 3.1), use a fresh sample of the unknown. On heating the solvent, remember that many organic liquids or their vapors are flammable and/or toxic!

Analysis by Groups

Group I   Chlorine- and Fluorine-Containing Plastics

Heat the sample in a test tube with an approximately 50% solution of sulfuric acid. An odor of acetic acid indicates copolymers of vinyl chloride and vinyl acetate.

If the result is negative, follow the procedure in Section 6.2.7 for behavior toward pyridine. The process of distinguishing plastics in this group on the basis of their solubility is long and the results are usually uncertain. In this group one also tests for fluorine-containing plastics, especially polytetrafluoroethylene and polytrifluorochloroethylene. No simple specific reaction is known for these materials. For their identification, in addition to their high density of $2.1-2.3\,g/cm^3$ and their complete insolubility at room temperature, one can use the test for fluorine (see Chapter 4). The key to identification of polytrifluorochloroethylene is the simultaneous positive result of a chlorine test. Polyvinyl fluoride and fluorine-containing elastomers are found less frequently, and it is not possible to identify them with simple tests.

Group II   Nitrogen-Containing Plastics

Diphenylamine test: Suspend 0.1 g diphenylamine in 30 ml water and then add carefully 100 ml concentrated sulfuric acid. (*Note*: The acid should be added slowly.) Add a drop of the fresh reagent to the plastic sample on a plate; a dark blue coloration indicates cellulose nitrate.

If the result is negative, test for bound formaldehyde. Heat a small sample of the plastic with 2 ml concentrated sulfuric acid and a few crystals of chromotropic acid for 10 min at 60–70 °C. A deep violet coloration indicates formaldehyde. Cellulose nitrate, polyvinyl acetate, polyvinylbutyral, and cellulose acetate give a red coloration; these materials, however, are not included in this part of the analytical procedure.

If the formaldehyde test is positive, heat a sample of the plastic with a 10% glycolic solution of potassium hydroxide (dissolve 10 g KOH in approximately 95 ml ethylene glycol). A smell of ammonia (confirmed by blue color of moist red litmus paper) indicates urea resins. Melamine resins do not liberate ammonia. However, they can be identified by the thiosulfate reaction and clearly distinguished from urea resins. For this purpose, heat a small amount of the sample in a test tube with a few drops of concentrated hydrochloric acid in an oil bath to 190–210 °C until congo red paper no longer turns blue. Cool the solution and add a few crystals of sodium thiosulfate. Cover the tube with a piece of congo red paper that has been moistened with 3% hydrogen peroxide, and heat in the bath to 160 °C. A blue color indicates melamine.

Thiourea resins are identified by the simultaneous presence of nitrogen and sulfur. (For identification of the individual elements, see Section 6.2.13.)

If the formaldehyde test is negative, cover a sample with nonaqueous sodium carbonate in a test tube and heat the tube until the material melts. The odor of ammonia indicates polyamides. If the vapors are acrid and neutral or lightly acidic to pH paper (sometimes also basic), this indicates urethanes. A sweet odor indicates polyacrylonitrile. Such vapors are clearly basic. (Test, see Section 6.2.4.) Group II tests are summarized in Figure 7.

Group III   Sulfur-Containing Polymers

In addition to polyalkylene sulfides, thiourea resins, and thiochlorinated polyethylene, this group includes sulfur-vulcanized natural and synthetic elastomers. Due to their rubberlike behavior, they will be discussed together with the identification reactions of elastomers in Section 6.2.18. Polysulfones and polyphenylene sulfides, which are used as engineering plastics, should also be considered in this group. If the sulfur-containing polymers, because of the simultaneous presence of nitrogen, have not been determined in Group II, similar to the thiourea resins,

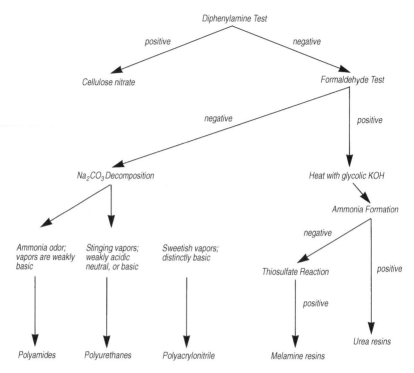

Fig. 7. Tests for group II, nitrogen-containing plastics.

see Sections 6.2.18 and 6.2.19 which deal with elastomers and with high temperature-resistant thermoplastics.

Polyalkylene sulfides (thioplastics) have a relatively high density $(1.3-1.6\,g/cm^3)$ and usually have a strong odor of hydrogen sulfide or mercaptans (like rotten eggs). The odor is especially strong on heating, and in this way they can be qualitatively identified.

Group IV   Plastics without Heteroatoms

The large group of plastics without heteroatoms can only be incompletely identified with this separation procedure. Place the sample in water. If it dissolves slowly, then it may be polyvinyl alcohol. (For specific identification, see Section 6.2.6.) If the plastic is insoluble in water, then check first for formaldehyde (Section 6.1.4). The only positive reaction in this group is given by phenol formaldehyde resins and polyoxymethylene (polyformaldehydes).

Next, test for phenols (see Section 6.1.3). They may result from phenol and cresol formaldehyde resins and also from epoxy resins or polycarbonates based on bisphenol A.

A further test for acetate (Section 6.2.5) makes it possible to identify polymers containing vinyl acetate as well as cellulose acetate or cellulose acetate butyrate (Section 6.2.16).

These tests, however, do not identify certain chemically very inert plastics such as polyethylene, polypropylene, polyisobutylene, polystyrene, polymethyl methacrylate, polyacrylates, polyethylene terephthalate, natural rubber, butadiene rubber, polyisoprene, and silicones. Their identification requires specific individual reactions, described in Chapter 6.

It is relatively simple to separate and characterize mixtures of polyvinylchloride, polyethylene and polystyrene which constitute a major part of all solid plastics waste. The procedure is as follows:

2 g of the mixture is stirred for 1 hour at room temperature in toluene. The insoluble residue is removed by filtration and dried at 80 °C (if possible in a drying closet). The filtrate contains polystyrene which can be separated from the solvent either by carefully evaporating the toluene, or by slowly adding the solution dropwise to about 300 ml methanol in order to precipitate the polystyrene. In order to confirm that the material is polystyrene, use the specific identification test for polystyrene described in Section 6.2.2. After separation of the polystyrene, the previously dried solid residue is treated with ca. 50 ml toluene for about 30 minutes at 80 °C on a waterbath. This treatment dissolves the polyethylene almost completely, whereas polyvinylchloride remains insoluble under these conditions. The contents of the flask are filtered while still hot and the solid residue is washed with heated toluene and then dried for an hour at 50 °C. To identify the material see Section 6.2.7. The polyethylene which was dissolved in the hot toluene precipitates on cooling of the solution to room temperature. It can be

recovered by filtration and identified according to the procedure given in Section 6.2.1.

An extensive compilation of physical properties, solubilities, pyrolysis behavior and some characteristic identifications of individual plastics are contained in the Plastics Identification Table of Hj. Saechtling (see Section 8.1).

6 Specific Identification Tests

6.1 General Identification Reactions

The reactions described in this section are useful as screening tests for certain groups of plastics, but also for testing for specific cleavage products, such as phenols or formaldehyde, from some plastics.

6.1.1 Liebermann-Storch-Morawski Reaction

Dissolve or suspend a few milligrams of the sample in 2 ml hot acetic anhydride. After cooling, add 3 drops of 50% sulfuric acid (from equal volumes of water and concentrated sulfuric acid). Watch the color reaction immediately and again after the sample has stood for 10 min. Heat it to 100 °C, using a water bath. This test is not specific but is often quite useful as an indicator (see Table 13).

6.1.2 Color Reaction with *p*-Dimethylaminobenzaldehyde

Heat 0.1–0.2 g of the sample in a test tube, and place the pyrolyzed product on a bare cotton plug. Drop the cotton in a 14% methanol solution of *para*-dimethylaminobenzaldehyde with a drop of concentrated hydrochloric acid. If polycarbonates are present, a deep blue color is produced. Polyamides show a bordeaux red color.

Table 13   Color Changes in the L-S-M Reaction

Material	Immediate	Color after 10 min	After Heating to about 100 °C
Phenolic resins	Reddish violet–pink	Brown	Brown-red
Polyvinyl alcohol	Colorless-yellowish	Colorless-yellowish	Brown-black
Polyvinyl acetate	Colorless-yellowish	Blue grey	Brown-black
Chlorinated rubber	Yellow brown	Yellow brown	Reddish to yellow brown
Epoxy resins	Colorless to yellow	Colorless to yellow	Colorless to yellow
Polyurethanes	Lemon yellow	Lemon yellow	Brown, green fluorescence

6.1.3 The Gibbs Indophenol Test

The Gibbs indophenol test is useful for the identification of phenol in phenolic resins and in substances that split off phenol or phenol derivatives on heating. Polycarbonates or epoxy resins as well as some HT-thermoplastics are examples of this. Heat a small sample for a maximum of 1 min in a pyrolysis tube and cover the opening of the tube with a piece of prepared filter paper. To prepare the paper, drench it in a saturated ether solution of 2,6-dibromoquinone-4-chlorimide and then air-dry it. After the pyrolysis, hold the paper over ammonia vapor or moisten it with 1–2 drops of dilute ammonia. A blue color indicates phenol (cresol, xylenol).

6.1.4 Formaldehyde Test

Heat a small sample of the plastic with 2 ml concentrated sulfuric acid and a few crystals of chromotropic acid for about 10 min at 60–70 °C. A strong violet color indicates the formation of formaldehyde. Cellulose acetate, cellulose nitrate, polyvinyl acetate, or polyvinylbutyral will yield a red color.

6.2 Specific Plastics

6.2.1 Polyolefins

Polyethylene and polypropylene are the polyolefins most commonly used as plastics. Polybutene-1 and poly-4-methylpentene-1 are less common. Also important are certain copolymers of ethylene and also polyisobutylene, which is used for gaskets. The simplest method of identification of these materials is by infrared spectroscopy (see Section 7.2). However, some information can also be obtained from the melting range (see also Section 3.3.3):

Polyolefin	Melting Range (°C)
Polyethylene (depending on density)	105–135
Polypropylene	160–170
Polybutene-1	120–135
Poly-4-methylpentene-1	above 240

A first indication that polyolefins are present can be obtained from a simple density measurement. Contrary to the behavior of other plastics, polyolefins without fillers float on water. The only other plastics that float on water are foamed plastics or those containing foaming agents.

The reaction of the pyrolysis vapors with mercury(II) oxide will differentiate between these materials. To do this, heat a dry sample of the plastic in the pyrolysis tube closed with a piece of prepared filter paper. To prepare the paper, drench it with a solution of 0.5 g yellow mercury(II) oxide in sulfuric acid (1.5 ml concentrated sulfuric acid added to 8 ml water, carefully). If the vapor gives a golden yellow spot, this indicates polyisobutylene, butyl rubber, and polypropylene (the latter only after a few minutes). Polyethylene does not react. Natural and nitrile rubber, as well as polybutadiene, yield a brown spot. Waxlike greases are the products in the pyrolysis of polyethylene and polypropylene. Polyethylene smells like paraffin, and polypropylene is slightly aromatic.

Polyethylene and polypropylene may also be differentiated by scratching the sample with your finger nail: whereas polyethylene will show scratch marks, polypropylene is scratch resistant.

6.2.2 Polystyrene

When polystyrene is heated in a dry test tube styrene monomer is formed which is easily identified by its typical odor.

Polystyrene and most styrene-containing copolymers can be identified by placing a small sample in a small test tube, adding a few drops of fuming nitric acid, and evaporating the acid without having the polymer decompose. The residue is then heated over a small flame for approximately 1 min. Fasten the test tube with its open end tilted slightly down and covered with a piece of filter paper. Prepare the paper by drenching it in a concentrated solution of 2,6-dibromoquinone-4-chlorimide in ether and then drying it in air. On moistening with a drop of dilute ammonia, the paper turns blue if polystyrene is present. If the sample still contains some free nitric acid, the test is affected and the paper turns brown, which may conceal the blue color. This identification is also useful for styrene-butadiene copolymers as well as for ABS (acrylonitrile-butadiene-styrene copolymers). The presence of acrylonitrile can be confirmed by a test for nitrogen.

6.2.3 Polymethyl Methacrylate

Polymethyl methacrylate plays an important role among the acrylates as an injection molding material as well as a glasslike material. For its identification, 0.5 g sample is heated in a test tube with approximately 0.5 g dry sand. On depolymerization, monomeric methyl methacrylate is obtained. This is captured at the opening of the test tube on a glass fiber plug. The methyl methacrylate monomer may be distilled from one test tube into another through a bent piece of glass tubing passing through a rubber stopper (Figure 8). Heat a sample of the monomer with a

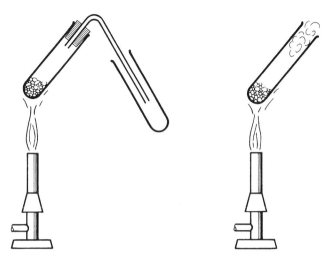

Fig. 8. Depolymerization in a test tube.

small amount of concentrated nitric acid (density $1.4 \, \text{g/cm}^3$) until a clear yellow solution is obtained. After cooling, dilute with approximately half its volume of water, then add a 5–10% sodium nitrite solution dropwise. Methyl methacrylate, which may be extracted with chloroform, is indicated by a blue-green color.

On pyrolysis, polyacrylates yield, in addition to monomeric esters, several strong-smelling decomposition products. The pyrolysates are either yellow or brown and acidic.

6.2.4 Polyacrylonitrile

Polyacrylonitrile is most often encountered as a fiber. It is also found in acrylonitrile-containing plastics copolymerized with styrene, butadiene, or methyl methacrylate. All such polymers contain nitrogen.

To identify acrylonitrile polymers, take a sample of the material and add a small amount of zinc dust and a few drops of 25% sulfuric acid (1 ml concentrated sulfuric acid added to 3 ml water, slowly). Heat this mixture in a porcelain crucible. Cover the crucible with filter paper moistened with the following reagent solution: Dissolve 2.86 g copper acetate in 1.0 liter water. Then dissolve 14 g benzidine in 100 ml acetic acid, and to 67.5 ml of this solution add 52.5 ml of water. Keep both the copper acetate and benzidine solutions in separate containers in the dark. Mix them in equal volumes just before use. The presence of acrylonitrile is indicated by a bluish spot on the filter paper.

The presence of acrylonitrile in copolymers can also be demonstrated by heating a sample of the dry material in a test tube and testing with indicator paper for the formation of HCN. Prepare the indicator paper by dissolving 0.3 g copper(II) acetate in 100 ml water. Impregnate strips of filter paper, then air-dry them. Just before use, dip the strips in a solution of 0.05 g benzidine in 100 ml 1 N acetic acid (prepared from equal parts of 2 N acetic acid and water). If HCN is formed and passes over the moist paper, the paper turns blue (Careful!).

When polyacrylonitrile is pyrolyzed, hydrogen cyanide (HCN) is formed and may be identified by means of the Prussian Blue reaction. 0.5 g of the polyacrylonitrile sample is completely pyrolyzed in a test tube and the pyrolysis vapors are introduced into 3 ml of dilute sodium hydroxide solution. After addition of about 1 ml ferrous sulfate solution, the mixture is brought to a boil and then reacted with a few drops of ferric chloride solution. After acidification with dilute hydrochloric acid, nitrile group containing polymers give rise to a characteristic blue color.

To distinguish between polyacrylonitrile and polyamides or polyurethanes, one dissolves a few mg of the sample in about 3 ml of dimethylformamide. After addition of about 3 ml of 60 % sodium hydroxide solution (carefully dissolve about 6 g NaOH in 10 ml water), the mixture is heated. An orange-red coloration is observed only if polyacrylonitrile is present.

6.2.5 Polyvinyl Acetate

Polymers containing vinyl acetate can be recognized by the fact that they produce acetic acid on thermal decomposition. Cellulose acetate behaves in a similar fashion. To test this, pyrolyze a small amount of sample and catch the vapors on water-moistened cotton. Then wash the cotton with some water and collect the liquid in a test tube. Add 3–4 drops of a 5% aqueous lanthanum nitrate solution, 1 drop 0.1 N iodine solution, and 1–2 drops concentrated ammonia. Polyvinyl acetate becomes deeply blue or almost black. Polyacrylate becomes reddish, polyvinyl acetate green to blue. As a further test, use the Liebermann-Storch-Morawski reaction (see Section 6.1.1). Polyvinyl acetate gives a purple-brown color on wetting with 0.01 N iodine/potassium iodate solution (0.1 N solution diluted to 10 times its volume). This color becomes stronger on washing with water.

6.2.6 Polyvinyl Alcohol

Saponification of polyvinyl acetate gives polyvinyl alcohol. The latter has no particular importance as a plastic raw material. The identification reactions will yield different results depending on the conversion during saponification. Highly saponified polyvinyl alcohols are insoluble in organic solvents but soluble in water and formamide. For the test involving the reaction with iodine, react 5 ml of the aqueous solution of polyvinyl alcohol with 2 drops of 0.1 N iodine-potassium iodide solution. Dilute this with water until the resulting color is only just recognizable. React 5 ml of this solution with as much borax as will fit on the tip of a spatula. Shake this and acidify it with 5 ml of concentrated hydrochloric acid. A strong green color, especially on the undissolved borax grains, indicates polyvinyl alcohol. The presence of starch and dextrin can interfere with this test.

6.2.7 Polymers Containing Chlorine

In addition to polyvinyl chloride (PVC), the chlorine-containing polymers and different copolymers of vinyl chloride are: polyvinylidene chloride, chloro-rubber, rubber hydrochloride, chlorinated polyolefins, polychloroprene, and polytrifluorochloroethylene. In addition to detecting chlorine with the Beilstein test (see Chapter 4), these polymers can be identified by using the color reaction with pyridine (see Table 14).

First the material must be freed of any plasticizers by extraction with ether. Alternatively, dissolve the sample in tetrahydrofuran, filter off possible undissolved components, and reprecipitate it by adding methanol. After extraction and drying at a maximum of 75 °C, react a small sample with 1 ml pyridine. Let it stand for a few minutes, then add 2–3 drops of a 5 % methanolic sodium hydroxide solution (1 g sodium hydroxide dissolved in 20 ml methanol). Note the color immediately, after 5 min, and again after 1 hr. For a more definitive test, boil a small amount of plasticizer-free material for 1 min with 1 ml pyridine. Divide the solution into two parts. Boil both portions again and then immediately add 2 drops of 5 % methanolic sodium hydroxide to one. Cool the other portion and then add to it 2 drops of methanolic sodium hydroxide. Observe the color immediately and after 5 min (See Table 14).

6.2.8 Polyoxymethylene

Polyoxymethylenes or polyacetals (polymers of formaldehyde or trioxane) produce formaldehyde on heating. The chromotropic acid test for formaldehyde is positive (see Section 6.1.4).

6.2.9 Polycarbonates

Almost all polycarbonates used as plastics contain bisphenol A. For positive identification, the color reaction with p-dimethyl-aminobenzaldehyde (see Section 6.1.2) or the Gibbs indophenol test (see Section 6.1.3) are used.

Polycarbonates are completely saponified in a few minutes on heating in 10 % alcoholic potassium hydroxide. Potassium carbonate precipitates during this reaction and can be filtered off. Acidify

Table 14 Color Reactions of Chlorine-Containing Plastics on Treating with Pyridine

Material	Boiled with Pyridine and Reagent Solution		Boiled with Pyridine; Cooled; Reagent Solution Added		Pyridine and Reagent Solution Added to Sample without Heating	
	Immediate	After 5 min	Immediate	After 5 min	Immediate	After 5 min
Polyvinyl chloride	Red-brown	Blood red, brown-red	Blood red, brown-red	Red-brown, black precipitate	Red-brown	Black-brown
Chlorinated PVC	Blood red, brown-red	Brown-red	Brown-red	Red-brown, black precipitate	Red-brown	Red-brown
Chlorinated rubber	Dark red-brown	Dark red-brown	Black-brown	Black-brown precipitate	Olive-brown	Olive-brown
Polychloroprene	White-cloudy	White-cloudy	Colorless	Colorless	White-cloudy	White-cloudy
Polyvinylidene chloride	Black-brown	Black-brown precipitate	Black-brown precipitate	Black-brown precipitate	Black-brown	Black-brown
PVC molding compound	Yellow	Black-brown precipitate	White-cloudy	White precipitate	Colorless	Colorless

the precipitate with dilute sulfuric acid, which will release carbon dioxide. On passing the gas into a barium hydroxide solution a white precipitate of barium carbonate forms.

6.2.10 Polyamides

The most important industrial polyamides are polyamides (Nylons) 6, 66, 610, 11, and 12. There are also a number of different copolymer amides that can be identified with simple means as polyamides (for example through the odor of burnt horn on exposure to a flame; see Section 3.3.2). However, complete identification is not always possible.

In some cases the melting point determination permits the distinction between the different polyamides:

Polyamide Type	Melting Range (°C)
Polyamide 6 (Nylon 6)	215–225
Polyamide 66 (Nylon 66)	250–260
Polyamide 610 (Nylon 610)	210–220
Polyamide 11 (Nylon 11)	180–190
Polyamide 12 (Nylon 12)	170–180

Polyamides can also be recognized through the color reaction with p-dimethylaminobenzaldehyde (see Section 6.1.2).

Polyamides can be readily identified by the acids formed on acid hydrolysis of the respective polyamides. For this purpose, heat 5 g of the sample with 50 ml concentrated hydrochloric acid using a reflux condenser (Figure 9). Continue reflux until the major part of the sample has dissolved. Then boil the solution with charcoal until the color disappears and filter it while it is hot. After cooling, filter off the precipitated acids and recrystallize them from a small amount of water. If no acids precipitate, extract the filtrate with ether. Evaporate the ether and recrystallize the residue from water. The acids have the following melting points:

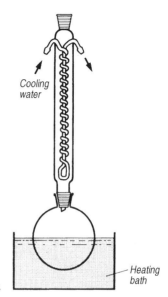

Fig. 9. Heating and boiling using reflux condenser.

Adipic acid (Nylon 66)	152 °C
Sebacic acid (Nylon 610)	133 °C
ε-Aminocaproic acid hydrochloride (Nylon 6)	123 °C
11-Aminoundecanoic acid (Nylon 11)	145 °C
12-Aminolauric acid (Nylon 12)	163 °C

Polyamides are easily differentiated from polyacrylonitrile by dissolution of the sample in dimethylformamide and subsequent addition of sodium hydroxide solution (see Section 6.2.4).

6.2.11 Polyurethanes

On pyrolysis, polyurethanes reform the isocyanates used in their synthesis to some extent. For their identification, heat a dry sample in a test tube, pass the resulting vapors over filter paper, and then moisten the filter paper with a 1% methanolic solution of 4-nitrobenzodiazoniumfluoroborate solution (Nitrazol CF extra, Hoechst AG). Depending on the type of isocyanate, the paper will turn yellow, reddish brown, or violet.

To distinguish between polyurethanes and polyacrylonitrile see Section 6.2.4.

6.2.12 Phenolics

Phenolic resins are made from phenol or phenol derivatives and formaldehyde. In many cases they also contain inorganic or organic fillers. After curing, the resins are insoluble in all the usual solvents, but they dissolve with decomposition in benzylamine. Phenolic resins may be identified in the Gibbs indophenol test (see Section 6.1.3). The bound formaldehyde may be identified with chromotropic acid (see Section 6.1.4).

6.2.13 Aminoplastics

The aminoplastics are condensation products of formaldehyde and urea, thiourea, melamine, or aniline. They are often filled with finely ground wood, stone, or other inorganic fillers and are used mainly as molded parts or laminates. All aminoplastics contain nitrogen and bound formaldehyde, which can be identified using chromotropic acid (see Section 6.1.4).

A specific identification test for urea is the enzymatic reaction with urease. 50 mg of powdered resin or 0.1 ml of the resin solution is carefully heated in a test tube with a Bunsen burner until all formaldehyde has been removed (check odor!). After cooling and neutralizing with 10% sodium hydroxide using phenolphthalein as indicator, 1 drop of 1 N sulfuric acid and 0.2 ml of a freshly prepared 10% urease solution are added. A moist piece of litmus paper is then attached to the upper rim of the test tube. After a short time the blue coloration of the indicator paper demonstrates the presence of ammonia which is formed only by urea-containing resins and not by melamine resins. Hexamethylene tetramine is the only substance that may interfere with this reaction.

Urea and thiourea resins can be identified by taking a few milligrams of the sample, adding 1 drop of hot concentrated hydrochloric acid (about 110 °C), and heating until dry. After cooling, add 1 drop of phenylhydrazine and heat the sample for 5 min in an oil bath at 195 °C. Cool it and add 3 drops of dilute ammonia (1 : 1 by volume) and 5 drops of a 10 % aqueous nickel sulfate solution. On shaking with chloroform, the solution becomes red to violet indicating the presence of urea or thiourea.

An additional test for sulfur (see Chapter 4) permits a distinction between urea and thiourea. Melamine resins can be recognized by pyrolysis. Heat a small sample with a few drops of concentrated hydrochloric acid. Use a pyrolysis tube and an oil bath at 190–200 °C. Cover the tube with congo red paper, heat it until the paper no longer turns blue, and cool. Then add a few crystals of sodium thiosulfate to the cooled residue. Cover the pyrolysis tube with congo red paper moistened with a 3 % hydrogen peroxide solution and heat it in the oil bath at 160 °C. In the presence of melamine, the paper becomes blue (urea resins do not react).

Aniline resins may be identified by pyrolytic decomposition. The addition of the gases produced to either sodium hypochlorite or calcium hypochlorite solution yields a red-violet or violet solution.

6.2.14 Epoxy Resins

There are no simple specific tests for the unconverted epoxy groups or the crosslinked units in hardened epoxy resins. Epoxy resins give a positive reaction for phenol according to the Gibbs indophenol test (see Section 6.1.4) (due to the presence of bisphenol A). In contrast to the phenolic resins, however, the formaldehyde test with chromotropic acid (see Section 6.1.4) is negative. All epoxy resins produce acetaldehyde during pyrolysis below 250 °C. Heat a sample of the material in a pyrolysis tube in an oil bath to 240 °C. Pass the vapors onto filter paper which has been moistened with a fresh aqueous solution of 5 % sodium nitroprussate and morpholine. A blue color indicates an epoxy resin.

An epoxy resin may also be recognized in the following way: Dissolve approximately 100 mg resin at room temperature in about 10 ml concentrated sulfuric acid. Then add about 1 ml concentrated nitric acid. After 5 min, top the solution carefully with 5% aqueous sodium hydroxide. In the presence of epoxy resins based on bisphenol A a cherry-red color will appear at the interface of the layers.

6.2.15 Polyesters

Unsaturated polyesters are produced in the form of dissolved resins in polymerizable monomers (usually styrene). They are also known in the form of molding resins or as hardened products. One should distinguish between them and the saturated aliphatic and aromatic polyesters. Among the latter are polyethylene terephthalate and polybutylene terephthalate.

The acidic components of most unsaturated polyesters are maleic, phthalic, sebacic, fumaric, or adipic acid. All of these can be directly identified.

Phthalic acid: Heat a small sample of the polymer with thymol (1 part sample to 3 parts thymol) and 5 drops of concentrated sulfuric acid for 10 min at 120–150 °C. After cooling, dissolve the sample in 50% ethanol and make the solution alkaline by adding 2 N sodium hydroxide. Phthalates produce a deep blue color.

Succinic acid may be identified by reacting a small amount of the resin (or 3–4 drops of the available resin solution) with about 1 g hydroquinone and 2 ml concentrated sulfuric acid. Heat this over a small flame to approximately 190 °C. Cool and dilute the sample with 25 ml water, and then shake it with about 50 ml toluene. The presence of succinic acid is confirmed when the solution turns red. Wash the toluene phase with water and react it with 0.1 N sodium hydroxide. A blue color results, however, phthalic acid, which may be present in these resins, could interfere.

Maleic acid resins give a wine red to olive brown color in the Liebermann-Storch-Morawski reaction (see Section 6.1.1).

Polyethylene terephthalate and *polybutylene terephthalate* are soluble in nitrobenzene. For their identification, pyrolyze a small sample of the solid plastic in a glass tube covered with filter paper. First drench the filter paper with a saturated solution of *o*-nitrobenzaldehyde in dilute sodium hydroxide. A blue-green color (indigo), which is stable against dilute hydrochloric acid, indicates terephthalic acid.

The absolute distinction between polyethylene terephthalate (PET) and polybutylene terephthalate (PBT) is difficult using simple methods. PET melts at 250–260 °C, PBT at about 220 °C. However, additives may cause deviations from these melting points.

PET and PBT can be identified by a white sublimate when these polymers are heated in a combustion tube.

6.2.16 Cellulose Derivatives

Cellulose acetate is the most well known plastic with a cellulose base. Others are cellulose acetobutyrate and cellulose propionate. Cellulose hydrate may be used as a vulcanized fiber. Cellulose may be identified fairly simply. Dissolve or suspend a sample in acetone, react it with 2–3 drops of a 2% solution of α-naphthol in ethanol, and carefully introduce a layer of concentrated sulfuric acid under this. At the phase boundary, a red to red-brown ring forms. In the presence of cellulose nitrate, a green ring forms. Sugars and lignin produce interference. For differentiation between cellulose acetate and cellulose acetobutyrate, it is usually sufficient to examine the vapors produced by dry heating of the sample. The acetate smells like acetic acid; the acetobutyrate smells of both acetic acid and butyric acid (like rancid butter).

For identification of cellulose acetates or propionates one can use the reaction with lanthanum nitrate. In this test one adds one or two drops of a 50% aqueous lanthanum nitrate solution and one drop of a 0.1 N iodine solution to a small amount of the polymer sample on a spot test plate. Then a drop of concentrated ammonia is added. If cellulose acetates are present one quickly observes a blue coloration; with cellulose propionate the coloration is brown.

Cellulose nitrates (Celluloid) may be recognized by the above reactions and by the sensitive diphenylamine test. Heat a sample with $0.5\,N$ aqueous potassium hydroxide ($2.8\,g$ potassium hydroxide in $100\,ml$ water) or $0.5\,N$ sodium hydroxide ($2.0\,g$ NaOH in $100\,ml$ water) for a few minutes and then acidify this with dilute sulfuric acid. Separate the supernatant liquid from the residue. Layer a solution of $10\,mg$ diphenylamine in $10\,ml$ concentrated sulfuric acid on top of that. A blue ring at the interface indicates cellulose nitrate. In order to identify nitrocellulose lacquers on cellophane, dissolve a few crystals of diphenylamine in $0.5\,ml$ concentrated sulfuric acid and add a few drops of this to the sample. A blue color is a positive test.

6.2.17 Silicones

Silicones are produced in the form of resins, oils, greases, and also as rubberlike elastic products. These materials also appear as processing aids in the manufacture of plastics, as impregnation compounds, coatings, separating materials, mold releases, etc. They can be identified because they contain the element silicon. To test for silicon, mix approximately $30\,mg$ of the sample with $100\,mg$ sodium carbonate and $10\,mg$ sodium peroxide. Heat this in a platinum or nickel crucible over a flame. Dissolve the melt in a few drops of water, boil it, and then add dilute nitric acid until the solution is neutral or slightly acidic. The identification of silicon then follows in the usual way with the addition of a few drops of ammonium molybdate. (See Chapter 4.)

6.2.18 Rubberlike Plastics

Although, strictly speaking, rubbers should not be classified as plastics, we would like to consider the most important types here since their areas of application overlap with those of plastics. Butyl rubber (polyisobutylene containing a few percent isoprene units) can be identified with mercury(II) oxide (see Section 6.2.1). Polybutadiene and polyisoprene contain double bonds, which can be identified using Wijs solution. This reagent solution is obtained by dissolving 6–7 ml of pure iodine monochloride in glacial acetic acid (up to 1 liter). The solution must be kept in the dark and has only a limited lifetime. To test the polymer, dissolve it in carbon tetrachloride or molten *p*-dichlorobenzene (melting point 50 °C) and react it dropwise with the reagent. Double bonds discolor the solution. This method is not specific for rubber but applies in general to all unsaturated polymers.

Use the Burchfield color reaction to differentiate between different types of rubber (see Table 15). Heat 0.5 g of the sample in a test tube. Pass the pyrolysis vapors into 1.5 ml of the reagent described below. Observe the color, dilute the solution with 5 ml of methanol and boil for 3 min.

Reagent. Dissolve 1 g *p*-dimethylaminobenzaldehyde and 0.01 g hydroquinone by heating them gently in 100 ml methanol. Then react solution with 5 ml concentrated hydrochloric acid and 10 ml

Table 15 Burchfield Color Reaction to Distinguish between Elastomers

Elastomer	On Contact of the Pyrolysis Vapors with Reagent	After Addition of Methanol and Boiling 3 min
None (blank test)	Yellowish	Yellowish
Natural rubber (polyisoprene)	Yellow-brown	Green-violet-blue
Polybutadiene	Light-green	Blue-green
Butyl rubber	Yellow	Yellow-brown to weakly violet-blue
Styrene-butadiene copolymers	Yellow-green	Green
Butadiene-acrylonitrile copolymers	Orange-red	Red to red-brown
Polychloroprene	Yellow-green	Yellowish green
Silicone rubber	Yellow	Yellow
Polyurethane elastomers	Yellow	Yellow

ethylene glycol. This reagent can be kept for several months in a brown bottle.

Rubber-like polymers include the thermoplastic elastomers (TPE) already mentioned in Chapter 1. They include mostly two-phase systems consisting of an elastic soft phase and a thermoplastic hard phase. The possible number of combinations is almost unlimited, which complicates their identification and nearly always necessitates expensive instrumental methods of analysis. In many cases the materials are block copolymers and, less frequently, blends. The following scheme provides an overview of the most important TPE types and can also be used as a guide for their qualitative analysis:

Block Copolymers	Thermoplastic Polyolefins (Blends, partly crosslinked)
Styrene-butadiene-styrene (SBS) Styrene-ethylene-butadiene-styrene (SEBS)	Polypropylene/ethylene-propylene terpolymers (PP-EPDM)
Ethylene-vinyl acetate/polyvinylidene chloride (EVA/PVDC) Thermoplastic polyurethane (TPU) Polyetherblock polyamide (PEBA) Copolyester, Polyetherester (TEEE)	Polypropylene/nitrile rubber (PP-NBR)

First, one tests for the components of the TPE according to the procedures described in the preceding sections (testing for double bonds in EPDM or butadiene components, see preceding paragraphs; polyolefins, see Section 6.2.1; styrene polymers, see Section 6.2.2; ethylene/vinyl acetate copolymers, see Section 6.2.5; polyurethanes, see Section 6.2.11; polyamides, see Section 6.2.10; terephthalic acid in polyetheresters, see Section 6.2.15).

6.2.19 High Temperature-Resistant (HT) Thermoplastics

The term HT-thermoplastics is used for polymers which in the absence of filler have a continuous-use temperature above approx.

200 °C. In contrast, standard plastics, such as PVC, polyethylene or polystyrene, have continuous-use temperatures of the order of 110 °C. In addition to their high temperature stability, HT-thermoplastics in general possess good resistance to chemicals and usually also low flammability. Among the most important HT-thermoplastics are: polyphenylene sulfides (PPS), polysulfones (PSU), polyether sulfones (PES), polyether imides (PEI), polyetherether ketones (PEEK) and polyarylates (PAR).

A first approach to differentiate the most important HT-thermoplastics is to examine their behavior in various solvents. Table 16 shows that PEEK and PES are completely insoluble in the most common solvents while PAR, PEI and PES dissolve easily in chlorinated solvents. Polyamide PA 6-3-T (polytrimethylhexamethylene terephthalate) does not dissolve in chlorinated solvents but is soluble in dimethyl formamide. Among the HT-polymers which are insoluble at room temperature, one can

Table 16   Dissolution Temperature of HT-Thermoplastics in °C

Solvent HF-Thermoplastic	Chloroform	Tetrahydrofuran	p-Xylene	Dichloromethane	Dimethylformamide	m-Cresol	Ethyl acetate	Trichlorobenzene	Methylethylketone	Toluene
PEEK	i	i	i	i	i	sw	i	i	i	i
PEEK (amorphous)	25	25	i	20	138	40	i	30	i	i
PEI	25	sw	i	20	45	58	i	110	i	i
PA 6-3-T	i	i	i	i	30	60	i	20	i	i
PAR	25	25	i	20	111	40	i	30	i	i
PESU	25	i	i	20	20	40	i	i	i	i
PPS	i	i	i	i	i	i	i	i	i	i

* i = insoluble; sw = swellable

Table 17   pH-Values of Pyrolysis Vapors from HT-Thermoplastics

Acidic (pH 0–5)	Neutral (pH 6–7)	Basic (pH 8–12)
PAR	PEEK	PEI
PESU	PEEK (amorphous)	PA 6-3-T
PPS		

recognize PPS by the fact that over approx. 210 °C it dissolves in chloronaphthalene and methoxynaphthalene.

Further information can be obtained from the pH-value of the pyrolysis vapors (see Table 17). With PAR one also notices a smell of phenol, which is an indication of the presence of phenolic components in the polymer chain. Pyrolysis of PES results in a stinging odor of sulfur dioxide; pyrolysis of PPS results in an odor of hydrogen sulfide; PEI and PA 6-3-T on heating smell like burnt horn.

The behavior of HT-thermoplastics when they burn can also be used for their identification (see Table 18).

Additional information can be obtained from certain special tests. Thus the Gibbs indophenol test is positive with PEEK, PAR and PEI (see Section 6.1.3). Using the color reaction with *p*-dimethylamino benzaldehyde (Section 6.1.2) one can differentiate HT-polyamides, such as PA 6-3-T, from polymers which develop phenolic decomposition products during pyrolysis. While with polyamides the red coloration obtained after addition of concentrated hydrochloric acid remains, with polycarbonates it turns blue.

6.2.20 Fibers

The following methods are used to identify polymers in textile fibers:

Flammability Tests: Burning wool smells like burnt horn, silk like burnt egg-white and cellulose fibers like burnt paper. Polyamide and polyester fibers melt before they burn, polyacrylonitrile fibers on burning leave a residue of hard, black spherical particles. On

Table 18   Burning Behavior of HT-Thermoplastics

Polymer	Flammability	Flame	Odor of Smoke	Indication
PEEK	extinguishes soon after removal from flame	slightly yellowish, slightly smoky	slight of Phenol	Phenol
PEEK (amorphous)	extinguishes immediately after removal from flame	yellow, foams	slight of Phenol	Phenol
PEI	extinguishes immediately after removal from flame	slightly yellowish, foams	burnt horn	Nitrogen (amides, imides)
PA 6-3-T	extinguishes immediately after removal from flame	blue flame border, very sooty	burnt horn	Nitrogen (amides)
PAR	extinguishes immediately after removal from flame	yellow, foams	of Phenol	Phenol
PESU	extinguishes immediately after removal from flame	yellow, foams	stinging of sulfur dioxide	Sulfur
PPS	extinguishes immediately after removal from flame	very sooty	strong of hydrogen sulfide	Sulfur

heating the dry fibers in a test tube, wool, silk and polyamides develop alkaline vapors, while cotton, bast fibers and regenerated cellulose (rayon) develop acidic vapors (test with moistened universal indicator paper).

Solubility Tests: Fibers of cellulose esters (e.g. cellulose acetate, cellulose nitrate) dissolve in acetone or chloroform, polyamide fibers in conc. formic acid, polyacrylonitrile fibers in cold, conc. nitric acid and in boiling dimethylformamide. Polyester fibers are soluble in 1,2-dichlorobenzene or nitrobenzene, wool dissolves in potassium hydroxide. Polyamide fibers can be differentiated by their different solubilities in 4.2 N hydrochloric acid: polyamide 66 (nylon 66) is soluble on heating, polyamide 6 (Nylon 6) dissolves at room temperature. (4.2 N HCl is prepared as follows: one carefully pours 35 ml of fuming (12.5 N) HCl into 65 ml of water).

These simple, rapid tests are not always reliable because with textiles consisting of a mixture of fibers, the results depend on the types and composition of the fibers. Further information may be obtained by means of infrared spectroscopy.

For the identification of wool and other animal hairs one can use the so-called plumbate reaction: formation of a black color resulting from the presence of sulfur in a water/alcohol solution of lead acetate heated to 80–90 °C.

6.3 Polymer Blends

In recent years mixed systems or blends of different polymers have been developed which are of increasing importance in the plastics industry. These materials, known as polymer blends or polymer alloys (see Table 3), are generally prepared by mixing two or more thermoplastics. They combine in an advantageous manner the properties of the thermoplastic components and in some cases the properties of the blend are superior to those of the individual components. (Polymer mixtures also result from the recycling of mixed plastics which have to be identified before they can be reused). Because of the large number of possible blend components and the fact that usually so-called compatibilizers of often rather complicated chemical composition are present, a complete analysis of polymer blends is not possible with simple methods. However, by means of some screening tests and selected special tests one can at least obtain qualitative information about the main components of such systems.

Solubility tests permit at least a tentative identification of the components also in polymer blends. Blends of ABS and polycarbonate are soluble in most polar solvents. Solubility in tetrahydrofuran and methyl ethyl ketone demonstrates the absence of polyolefins in such blends and the presence of aromatic polyesters or of polyamides can also be excluded. On the other hand, they may contain such generally highly soluble polymers as polystyrene, PVC, ABS, or polymethacrylates. However, blends that contain polybutylene terephthalate or polyethylene terephthalate do not dissolve in the usual solvents but require *m*-cresol, which can be a clear indication that aromatic polyesters are present. Polyolefins dissolve at high temperatures, at least 110 °C, in toluene and *p*-xylene and this behavior is characteristic of blends that contain polyethylene or polypropylene.

Pyrolysis experiments are useful for a further analysis of polymer blends. One can distinguish different groups of polymers by the pH-value of the resulting pyrolysis vapors. The formation of neutral pyrolysis vapors indicates that the blend does not contain any heteroelements and, therefore, usually belongs to the polyolefin group. Slightly acidic vapors (pH 3–5) can result from the formation of phenols or weak acids, such as terephthalic acid, which indicates that the blend may contain polycarbonates or aromatic polyesters. PVC-containing blends develop strongly acidic vapors due to the formation of hydrochloric acid which can be detected by its stinging odor (careful!) or the formation of a white fog of ammonium chloride when the vapors are passed over steaming ammonia.

The flammability behavior of some polymer blends is summarized in Table 19. Here we can also obtain some useful information: thus if the material burns readily while melting slowly and emits a definite paraffinic odor, this points to a polyolefin blend. Highly sooty flames are a definite indication that aromatic structures are present, while the odor of burnt horn indicates nitrogen-containing components. The identification of PVC in blends is relatively easy because generally a stinging smell of hydrochloric acid develops while the material usually burns very poorly. When polycarbonates are present in the blend, a typical odor of phenol is noticed in most cases.

When testing polyolefins with mercury(II) oxide (see Section 6.2.1) in most cases a yellow color develops more or less quickly.

Table 19 Flammability Behavior of Polymer Blends

Blend	Flammability	Flame	Odor of Smoke	Indication
PVC-Blends	extinguishes immediately after removal from flame	green flame border	stinging odor of hydrochloric acid	Chlorine
ABS/PC	continues to burn after removal from flame	bright flame, very sooty	weak odor of styrene and burnt rubber	Styrene copolymers
PBT/PC	continues to burn after removal from flame	bright flame, very sooty	sweetish, stinging	Polycarbonate
PBT/PET	continues to burn after removal from flame	bright flame, very sooty	sweetish, scratchy	Terephthalate
PE/PP	continues to burn after removal from flame	slightly yellowish, blue flame border	paraffinic odor, weakly like burnt rubber	Polyolefins

With blends of polyolefins no significant color differentiation sufficient for identification occurs, however if the polyethylene content of the blend is very high no or only a slight coloration is seen because polyethylene does not react with mercury(II) oxide.

Identification of polymer blends by their softening and melting behavior is not possible with simple methods. However, glass transition temperatures and melting temperatures can be readily determined by means of differential thermal analysis (DTA). Because most polymers are not miscible, the characteristic values for the individual components of the blend can be observed in parallel and more or less unchanged. Finally it should be pointed out that IR-spectroscopy can also be utilized to identify the individual components in polymer blends. However, in many cases the absorption bands of the components overlap and, therefore, it is necessary to compare the blend with known samples which is difficult and thus not a suitable approach to the identification of unknown materials.

6.4 Detection of Metals in Polyvinyl Chloride (PVC)

Processed polyvinyl chloride (PVC) almost always contains some metal-containing heat stabilizers, such as lead salts, metal soaps, and organotin compounds. The metal soaps consist of different combinations, mainly of barium, cadmium, zinc and calcium carboxylates of longer chain fatty acids. Of practical importance are mainly the calcium/zinc, barium/zinc, barium/cadmium and barium/cadmium/lead systems, whereas lead and tin stabilizers generally are not combined with other metals. Although in recent years the use of lead- and cadmium-containing stabilizers is greatly reduced, one still finds these metals in older materials and in old PVC that is supplied to recyclers. For qualitative identification of these metals in processed PVC they have to be converted into a water-soluble form. Once the material has been brought into solution, it is possible by means of test strips to directly identify lead, calcium, zinc or tin. For barium and cadmium, because test strips are not commercially available, the following spot tests are recommended.

To bring the material into solution approx. 1 g of PVC is dissolved in 20 ml of tetrahydrofuran by stirring or shaking in a 50 ml Erlenmayer flask. If the solution is cloudy, which indicates the presence of filler, a few drops of 65 % nitric acid are added and the solution is then filtered. The clear solution is added slowly to 60 ml methanol in order to precipitate the polymer. After filtering out the precipitate, it is evaporated to dryness, best by using a rotating evaporator but it can also be dried under vacuum or in a well-ventilated hood (Careful! Tetrahydrofuran is flammable!). The residue is then dissolved in a small amount of water and a few drops of dilute nitric acid. This solution can be used directly for the identification of the various metals by means of test strips available from chemical supply companies.

Test for Cadmium: The following reagent is used: 0.5 g 2,2-bipyridyl and 0.15 g iron(II) sulfate ($FeSO_4 \cdot 7H_2O$) are dissolved in 50 ml water and treated with 10 g potassium iodide. After shaking for 30 min the solution is filtered. If the solution, which normally is stable, turns cloudy, it should be filtered before use. Procedure: one drop of the slightly acidic, neutral or slightly basic sample solution is deposited on a piece of filter paper and immediately, before the drop is absorbed by the filter paper, a drop of the reagent is added. A red spot or ring indicates the presence of cadmium.

Test for barium: A drop of the neutral or slightly acidic sample solution is deposited on a piece of filter paper and then treated with a drop of a 0.2 % aqueous solution of sodium rhodizonate. A reddish-brown spot indicates the presence of barium.

The most important metal-containing heat stabilizers in PVC can also be identified through IR-spectroscopy (see Section 7.2). Characteristic absorption bands for the salts of carboxylic acids are the bands of the ionized carboxyl group in the region from 1590 to $1490\,cm^{-1}$ and from 1410 to $1370\,cm^{-1}$. Since the exact position of the absorption bands depends mainly on the metal counterions of the carboxyl group, the IR-spectra provide a first identification of these metals. Tin stabilizers also show characteristic bands in these regions of the IR-spectrum. To determine the spectra one uses pressed pellets made by finely grinding the sample material with potassium bromide. The identification of the metal becomes even more certain if an FT-IR (Fourier Transform–Infrared) instrument is available. With such an instrument it is possible to subtract the spectrum of an additive-free PVC sample from that of a sample containing stabilizer with the result that the stabilizer peaks in the resulting spectrum are observed much more clearly. With highly filled samples or those containing plasticizers there may be some interferences.

Tables 20 and 21 show the typical IR-absorption bands of metal-containing stabilizers used in PVC.

Table 20 Typical IR-Absorption Bands of the Carboxyl Group of Metal-Containing Stabilizers in PVC

Cation	1. Absorption Band (strong) (cm^{-1})	2. Absorption Band (medium) (cm^{-1})
Calcium	1575	1538
Barium	1515	1410
Zinc	1540	1400
Cadmium	1535	1405
Lead	1535	1508
Tin	1575	1640/1602/1564

Table 21 IR-Absorption Bands of the Most Important Organic Residues of PVC Stabilizers

Anionic Residue	Absorption Bands (cm^{-1})
Unbranched fatty acids	720
Phthalic acid	870–650
Maleates	870–860
Mercaptides	1440–1410 and 1620–1220

7. Advanced Analytical Methods

7.1 Overview

Previous chapters have described simple methods of identification and, generally, these are sufficient in order to assign an unknown sample to a certain group of plastics. Of course, with these simple methods, especially in dealing with plastics of complicated composition, one can only obtain some qualitative information. In order to obtain more detailed information it is necessary to use more advanced methods of analysis (Table 22) which can only be carried out by specially trained personnel.

Plastics analysis is not only concerned with the identification of polymeric materials but also with the qualitative and quantitative determination of the additives and chemical aids which have been added to the plastic before or during processing. In a wider sense, plastics analysis also encompasses the determination of the properties of the end product. Theses properties depend not only on the starting materials but also to a considerable extent on the processing conditions. As the focus of plastics analysis widens it is also constantly gaining in importance for quality control and quality assurance purposes.

Similarly, the increasing efforts in recent years to recycle used plastics and plastics waste involve a number of analytical problems. Plastics waste containing a mixture of plastics can often be sorted using simple methods to identify the different plastics. It is still difficult, however, to automatically identify the

Table 22   Characterization Methods for Plastics

Structure of the individual Macromolecule

Chemical composition	Spectroscopic methods: IR, FT-IR, Raman, UV, NMR (H, C), atomic absorption spectroscopy Pyrolysis—gas chromatography —mass spectrometry
Steric structure	NMR, FT-IR, X-ray scattering
Molecular size/ molecular weight	Number average: vapor pressure and membrane osmometry, end group analysis
	Weight average: light scattering, ultracentrifuge
	Coil dimension: light scattering, small-angle x-ray scattering, neuton small-angle diffraction, viscometry in solution and melt
	Measured quantities: limiting viscosity, Fikentscher K-value or Mark-Houwink K and a, rheometry, flow curves, melt index gel permeation chromatography(GPC) Turbidity titration
Branching	Viscosity, light scattering, FT-IR
Chemical and steric non-homogeneity, sequence length	Chemical degradation, NMR, chromatography

Molecular Structure of the Polymer

Distribution of the chemical composition	Fractionation, HPLC, GPC-IR, -UV, -NMR
Sequence length distribution	Degradation of fractions, NMR
Steric non-homogeneity	Fractionation, NMR
Molecular size distribution (non-homogeneity)	GPC, precipitation fractionation, ultracentrifuge

Table 22   Characterization Methods for Plastics (continued)

Molecular Structure of the Polymer (continued)

Molecular motion	Mechanical and dielectric relaxation spectroscopy, NMR, dynamic spectroscopy and diffraction (IR, Raman, birefringence, x-ray, light scattering), quasielastic neutron scattering, rheology
Crystallinity and melting temperature	IR-spectroscopy, solid-state NMR, small-angle light scattering, large-angle x-ray scattering, light- and neutron scattering, density measurements, dilatometry, differential thermal analysis, polarization microscopy, electron microscopy

Supermolecular Structure

Particle size	Electron microscopy, light scattering, ultracentrifugal sedimentation, light microscopy
Crosslinking	Modulus of elasticity, degree of swelling
Glass transition	Dynamic-mechanical analysis, temperature dependence of the dielectric constant, viscosity, index of refraction and other properties, differential thermal analysis, dilatometry
Orientation	Polarization microscopy, IR-dichroism,

different plastics and to sort them at the same time. With used plastics and plastics which have been exposed to the environment for years, the analytical problems are even more difficult, for example if one wants to determine whether an aged plastic satisfies the quality demands for reuse of the material. As a result, in addition to the classical so-called wet chemical methods which have been used for a long time and some types of chromatography, more and more physical methods are used in the analytical laboratory. Most important among these are the different chromatographic, spectroscopic and thermoanalytical techniques. While it is not possible within the framework of this small book to discuss the large number of modern and mostly instrumental techniques, Table 22 presents an overview of the properties important for the characterization of a plastic and the methods used for their determination. Clearly, such characterization methods require considerable investment in instrumentation and qualified operators. These conditions are met by academic and industrial research laboratories and by government and private testing laboratories.

7.2 Infrared Spectroscopy

Of the analytical methods mentioned in Section 7.1, infrared spectroscopy, because of the availability of affordable and user-friendly instruments, is now used routinely in analytical laboratories. For many years it has been a classical method for the analysis of plastics and additives. There exist extensive collections and databanks of IR-spectra which are of help for qualitative identification purposes but, after appropriate calibration, also for quantitative determinations. Problems may be encountered when samples are not sufficiently transparent to the IR-radiation, as is the case with highly filled and cross-linked samples. In such cases reflection methods and photo-acoustic spectroscopy have been employed successfully, but these methods, because of their complexity will not be discussed here.

Recently, near-infrared spectroscopy (NIR), which utilizes the region of the infrared spectrum between wavenumbers of about $4000\,\mathrm{cm}^{-1}$ to $10\,000\,\mathrm{cm}^{-1}$, has achieved increasing prominence for the analysis of mixtures of plastics, in particular because of its ability to directly analyze solid materials. NIR permits the examination of incoming plastics from the supplier or of plastics wastes in a matter of seconds.

Infrared spectroscopy is based on the fact that light in the region of wavelengths between 750 μm and 1 mm causes molecules or sections of molecules to vibrate. Theses vibrations resulting from the absorption of the incident light appear in the infrared spectrum as absorption bands. The energy absorbed from infrared light by certain chemical bonds or groups at a certain wavelength leads to reduced transparency (transmittance) which is normally plotted as a function of the wavenumber (in cm^{-1}). The resulting infrared spectra are often sufficient for the identification of plastics, usually by comparing the spectrum of the unknown sample with those of known plastics found in a spectral collection (e.g. Hummel/Scholl: "Atlas of Polymer and Plastics Analysis, 3rd. Edition", Hanser/Wiley-VCH).

IR-spectra are obtained from samples in film form, provided the film is thinner than approx. 50 μm. Thicker samples or resin granules are heated above their softening temperature and then pressed to form films thin enough to be used directly for IR-spectroscopy. Films can also be obtained by casting from solutions of the plastic. A few drops of the solution are placed on a potassium bromide disc and after evaporation of the solvent the IR-spectrum can be observed directly from the disc because KBr does not show any absorption in the infrared, i.e. it is completely transparent. One has to make sure that evaporation of the solvent is complete in order to avoid interference through absorption bands due to the solvent. For this reason evaporation is completed by warming of the KBr disc with the sample on a hot plate or with a hairdryer or in a dessicator. If for some reason films cannot be prepared, the plastic can also be finely ground, mixed with KBr powder and then pressed into a disc. Usually 1 mg of the sample is mixed with approx. 100 mg KBr, thus only very small amounts of sample are required, however, it is important that the finely ground sample is well mixed with the KBr in a mortar or a mechanical mixer before pressing. With rubber-like polymers, e.g. natural rubber and synthetic elastomers, grinding at room temperature may not be possible and grinding may have to be carried out at low temperature using dry ice or liquid nitrogen. Once grinding and mixing with KBr is complete, the mixture can be brought to room temperature and pressed in a special press into a tablet which can be placed directly into the spectrometer. It should be noted that KBr is hygroscopic and, therefore, should always be used after careful drying to avoid interference in the absorption spectrum by water bands at about 3300 and 1640cm^{-1}.

Section 7.3 shows the infrared spectra of the most important polymers discussed in this book. IR-spectra of many other plastics and additives can be found in various extensive collections and databanks available also from the manufacturers of IR-spectrometers. It should be kept in mind that the infrared absorption bands of additives, such as plasticizers, antioxidants, fillers and pigments will interfere with those of the pure plastic and that, therefore, for identifaction of the plastic these substances will first have to be removed or other additional analytical methods will have to be employed. Table 23 coordinates various polymers with their IR-absorption bands. In addition one should compare the spectrum of any unknown sample with spectra of known polymers either from personally prepared materials or from available spectra collections. It should be noted that spectra shown in various publications are not always recorded using the same units for the vertical and horizontal axes and, therefore, may have a completely different appearance.

7.3 IR-Spectra

Table 23   IR-Bands of Various Polymers

A. Carbonyl bands present (1710–1780)

Aromatic bands (at 1600 and 1500 cm^{-1})	
Present	Absent
Polyurethanes (3330, 1695, 1540, 1220)	Polyvinyl acetate (1370, 1240, 1025)
Polyester urethanes (3330, 1700, 1540, 1410, 1220)	Polymethyl methacrylate (1485–1450[d], 1265–1240[d], 1190–1150[d])
Polyether urethanes (3330, 1700, 1540, 1410, 1310, 1220, 1110)	Polyacrylates (1250, 1160[b])
Polyesters of terephthalic acid (1260, 1110, 1020, 980, 880, 725)	Cellulose esters (1230, 1170–1050[b])

B. Carbonyl bands absent

Aromatic bands (at 1600 and 1500cm)	
Present	Absent
Phenolic resins (3330, 1220, 910–670)	Polyethylene (1470, 1380, 730–720)
Bisphenol epoxides (1235, 1180, 1040, 830)	Polypropylene (1470, 1380, 1160, 1000, 970)
Polystyrene (1450, 760, 700)	Polyacrylonitrile (2940, 2240, 1450, 1070)
ABS (2240, 970, 760, 700)	Aliphatic polyamides and urea resins (3300, 1640, 1540)
Styrene-butadiene copolymers (1450, 970, 760, 700)	Cellulose nitrate (1640, 1280, 1160–1060–1000[b], 834)
SAN (2240, 1450, 760, 700)	Melamine resins (3420, 1540[b], 813)
Polysulfones (with bisphenol A) (1320[d], 1250, 1165[d], 875–850–835, 560)	Polyvinyl chloride (1430, 1335, 1250, 1100, 960, 690, 630)
Polyphenylene oxide (1310, 1190, 1020, 855)	Polyvinylidene chloride (1400, 1350, 1050[d], 660, 600, 430)
Polyphenylene sulfide (1390, 1095, 1075, 1010, 820, 560)	Polytetrafluoroethylene (1220–1150[d])
Aromatic polyamides (3300, 1640, 1540)	Polytrichlorofluoroethylene (1250–1110, 1000–910)
	Aliphatic polysiloxanes (1265, 1110–1000, 800)
	Cellulose/cellophane (3450[b], 1160–1000[b])
	Polyoxymethylene (1240, 1090, 935[b])
	Polyethylene oxide (1470, 1360, 1275, 1140, 1110, 1060, 950[d], 840)

b: broad, d: double band

List of Spectra

1. Polyethylene, high density (HDPE)
2. Polypropylene (PP)
3. Polystyrene (PS)
4. Styrene-acrylonitrile (SAN)
5. Acrylonitrile-butadiene-styrene (ABS)
6. Polymethyl methacrylate (PMMA)
7. Polyvinyl chloride (PVC)
8. Polyoxymethylene (POM)
9. Polycarbonate (PC)
10. Polyamide 6 (Nylon 6) (PA 6)
11. Polyamide 66 (Nylon 66) (PA 66)
12. Polyesterurethane
13. Polyetherurethane
14. Phenolic resin (PF)
15. Melamine resin (MF)
16. Epoxy resin (EP)
17. Polyethylene terephthalate (PET)
18. Polybutylene terephthalate (PBT)
19. Cellulose acetate (CA)
20. Polydimethyl siloxane
21. Polytetrafluoroethylene (PTFE)

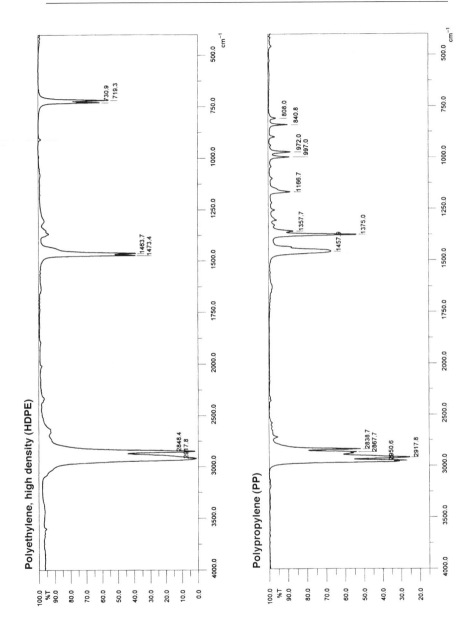

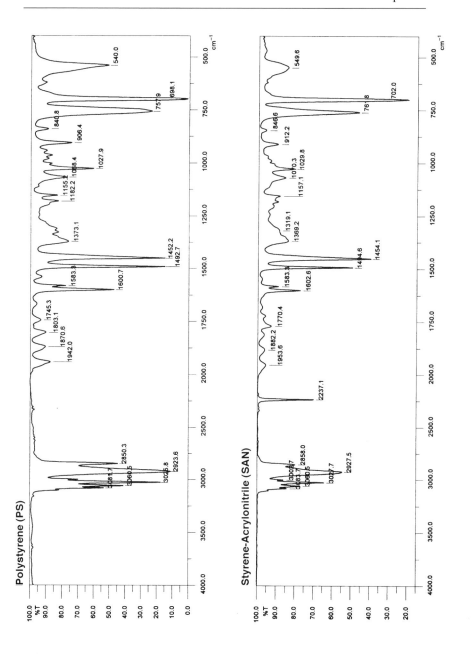

Polystyrene (PS)

Styrene-Acrylonitrile (SAN)

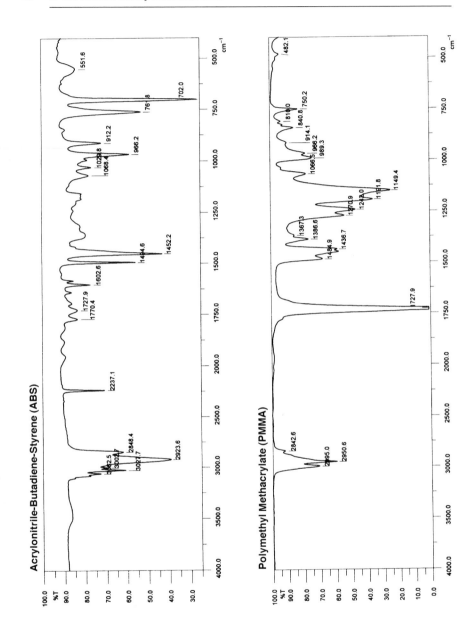

Acrylonitrile-Butadiene-Styrene (ABS)

Polymethyl Methacrylate (PMMA)

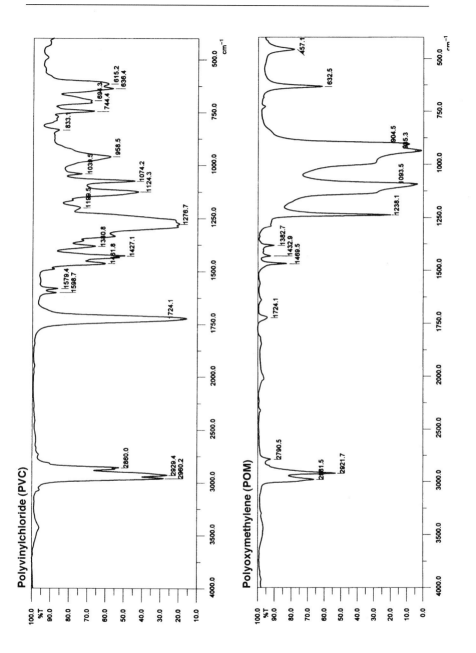

Polyvinylchloride (PVC)

Polyoxymethylene (POM)

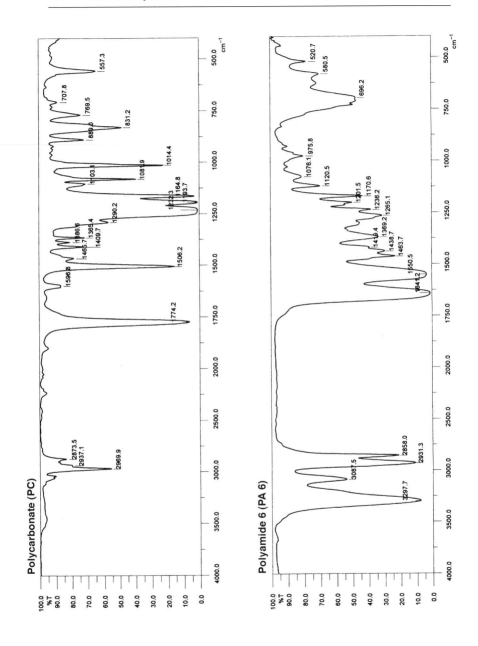

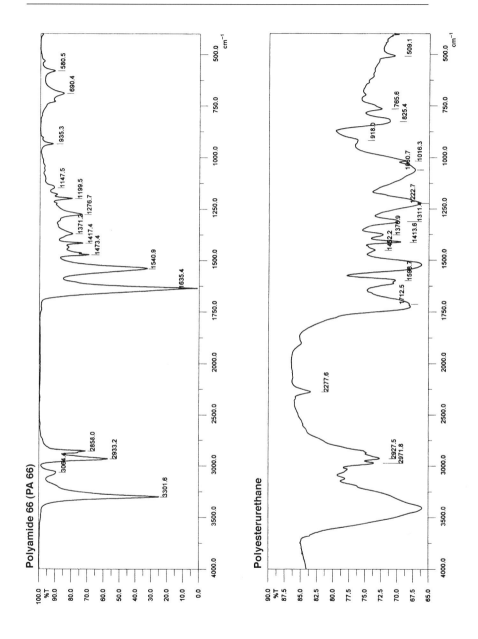

Polyamide 66 (PA 66)

Polyesterurethane

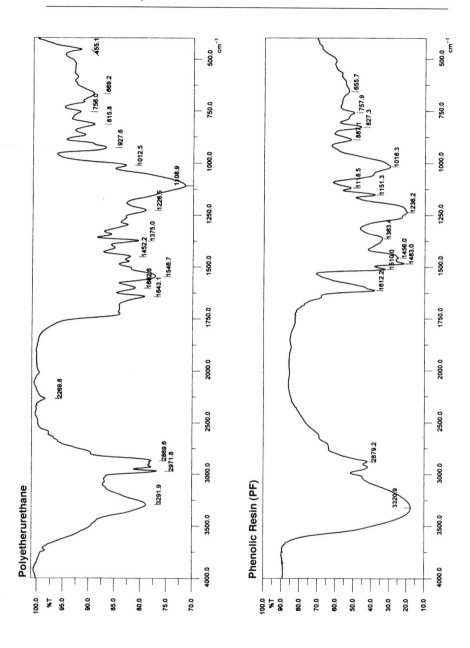

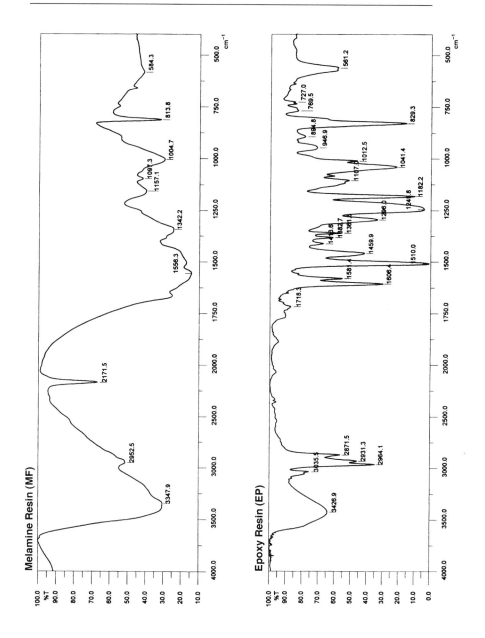

Melamine Resin (MF)

Epoxy Resin (EP)

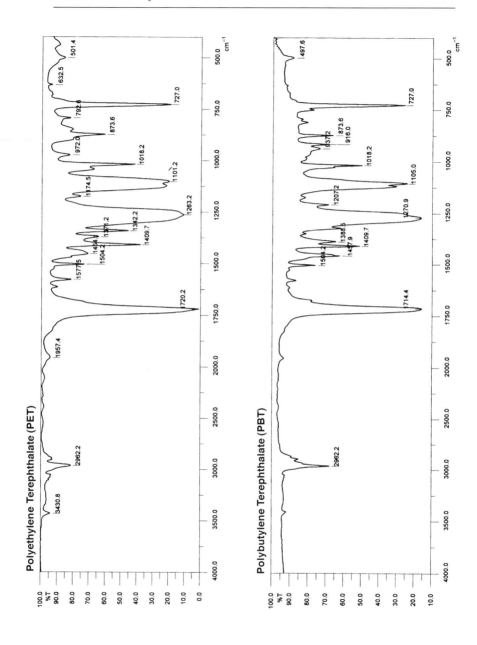

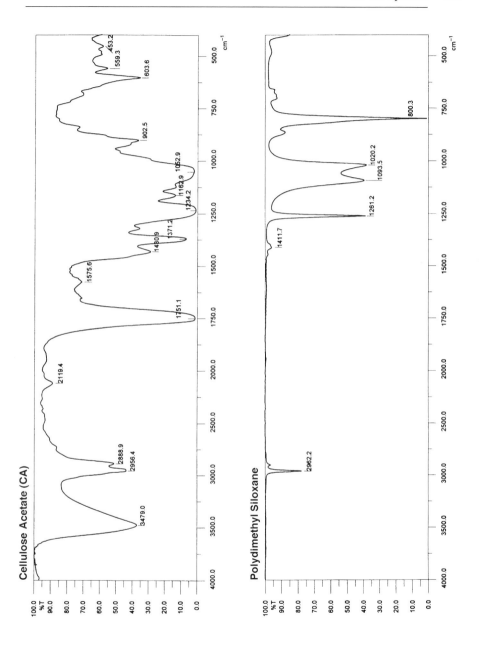

Cellulose Acetate (CA)

Polydimethyl Siloxane

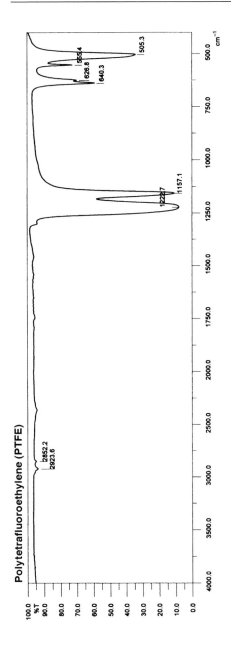

8 Appendix

8.1 Plastics Identification Table (after Hj. Saechtling) (Table 24)

Physical Properties

Standard Abbreviation (ISO 1043/ASTM 1600)	Name	Density unfilled g/cm³	transparent thin films	transparent, clear	hazy to opaque	usually contains fillers	Gasoline	Toluene	Methylene chloride	Diethylether	Acetone	Ethylacetate	Ethylalcohol	Water
1 Polyolefins														
PE	Polyethylene (chlorinated PE see group 3)	soft to hard ≥0.92 ≤0.96	+		+		i/sw / i	sw / i/sw	i / i/sw	i/sw / i	i/sw / i	i/sw / i	i / i	i / i
PP	Polypropylene	0.905	+		+		i/sw	i/sw	i/sw	i	i	i	i	i
PB	Polybutene-1	0.915			+		sw	i/sw	i	i/sw	i	i/sw	i	i
PIB	Polyisobutylene	(0.93)		+		+	s	s	s	sw	i	i	i	i
PMP	Poly-4-methylpentene-1	0.83			+		sw	sw	i	i	i	sw	i	i
2 Styrene Polymers														
PS	Polystyrene (pure)	1.05	+	+			sw/s	s	s	i/sw	s	s	i	i
SB	High impact polystyrene (with polybutadiene)	1.05			+	+	sw/s	s	s	s	s	s	i	i
SAN	Styrene-acrylonitrile copolymer	1.08	+	+			i	s	s	s	s	s	i	i
ABS	Acrylonitrile-butadiene-styrene copolymer	1.06			+		sw	s	s	s	s	s	sw	i
ASA	Acrylonitrile-styrene-acrylate copolymer	1.07			+		sw	s	s	sw	s	s	sw	i

Solubility in Cold Solvents (ca. 20 °C): s = soluble, sw = swellable, i = insoluble

Rate of dissolution depends on type of copolymer

Pyrolysis and Identification Tests

Elastic Behavior — leathery or rubbery, soft	flexible, resilient	hard	Sample slowly heated in pyrolysis tube (m = melts, d = decomposes; al = alkaline, n = neutral, ac = acidic, sac = strongly acidic)	Reactions of vapors given off	Ignition with small flame (0 = hardly ignitable, I = burns in flame, extinguishes in absence of flame, II = continues to burn after ignition, III = burns vigorously, fulminates)	Odor of vapors given off on heating in pyrolysis tube or after ignition and extinction	Characteristic elements (N, Cl, F, S, Si), individual identification tests (Compare Chapter 4 "Testing for Heteroatoms" and Chapter 6 "Specific Identification Tests")	Plastic Materials (arranged in chemical groups) — Name	Standard Abbreviation (ISO 1043/ASTM 1600)
								1 Polyolefins	
							Different melting ranges:		
+			becomes clear, m, d, vapors are barely visible	n	yellow with blue center, burning droplets fall off	slight paraffin-like odor, PP and PB with different flavour	105–120 °C	Polyethylene soft to hard	PE
+	+						125–130 °C		
	+						165–170 °C	Polypropylene	PP
	+	+					130–140 °C	Polybutene-1	PB
+			m, vaporizes, gases can be ignited	n	II	yellow, burns quietly	paraffin- and rubber-like	Polyisobutylene	PIB
	+		m, d, vaporizes, white smoke	n	II	yellow with blue center, drips	slightly paraffin-like	245 °C	Poly-4-methylpentene-1	PMP
								2 Styrene Polymers	
	+		m and vaporizes	n	II		characteristic of city gas	On breaking by hand: brittle fracture	Polystyrene (pure)	PS
	+	+	m, yellowish, d	n	II	flickers, yellow, bright	like PS + rubber	white fracture	High impact polystyrene (with polybutadiene)	SB
	+	+	m, yellow, d	al	II	very sooty	similar to PS, irritating	N, brittle fracture	Styrene-acrylonitrile copolymer	SAN
	+	+	d, turns black	n (ac)	II		like PS + cinnamon	N, white fracture	Acrylonitrile-butadiene-styrene copolymer	ABS
		+	m, d, black residue	(ac)	II		like PS + pepper		Acrylonitrile-styrene-acrylate copolymer	ASA

Physical Properties (continued)

Standard Abbreviation (ISO 1043/ASTM 1600)	Name	Density unfilled g/cm³	transparent thin films	transparent, clear	hazy to opaque	usually contains fillers	Gasoline	Toluene	Methylene chloride	Diethylether	Acetone	Ethylacetate	Ethylalcohol	Water
						solid products	s = soluble / sw = swellable / i = insoluble							
3 Halogen-containing Homopolymers														
PVC	Polyvinylchloride, ca. 55% Cl	1.39	+	+			i	i/sw	i/sw	i	i/s	i/sw	i	i
	Copolymer with VAC (or similar)	1.35					copolymers sw/s easier than PVC							
PVCC	High temp. resistant, 60–67% Cl	~1.5	+	(+)			i	i/sw	i	i	i/sw	i		i
PVC-HI	Impact resistant to high impact resistant:													
	made elastic with EVAC (or similar)	1.2–1.35	+		+		i/sw	i/sw	sw/s	i/sw	sw	i/sw	i	i
	made elastic with chlorinated PE	1.3–1.35	+		+		i	sw	sw	i	i	i	i	i
PEC	Chlorinated PE (pure homopolymer)	1.1–1.3			+		sw	between PE and PVC depending on Cl-content						
PVC-P	Plasticized (properties depending on plasticizer)	1.2–1.35	+	+	+		i	sw	sw	sw	sw	sw	sw	i
							Plasticizer (usually) removed by Diethyl ether							
PTFE PFEP PFA ETFE	Polytetrafluoroethylene; PTFE-like molding materials	} 2.0–2.3 / 1.7			+		i	i	i	i	i	i	i	i
							PFA, ETFE sw in hot CCl₄ (or similar)							
CTFE	Polytrifluorochloroethylene	2.1		+	+		i	i	i	i	i	i	i	i
PVDF	Polyvinylidene fluoride	1.7–1.8	+	(+)	+		i	i	sw	i/sw	sw		i	i

Pyrolysis and Identification Tests (continued)

leathery or rubbery, soft	flexible, resilient	hard	Sample slowly heated in pyrolysis tube (m = melts, d = decomposes)	Reactions of vapors given off (al = alkaline, n = neutral, ac = acidic, sac = strongly acidic)	Ignition with small flame (0 = hardly ignitable, I = burns in flame, extinguishes in absence of flame, II = continues to burn after ignition, III = burns vigorously, fulminates)	Odor of vapors given off on heating in pyrolysis tube or after ignition and extinction	Characteristic elements (N, Cl, F, S, Si), individual identification tests	Plastic Materials (arranged in chemical groups) — Name	Standard Abbreviation (ISO 1043/ASTM 1600)
								3 Halogen-containing Homopolymers	
	+			sac	I			Polyvinylchloride, 55% Cl; Copolymer with VAC (or similar)	PVC
	+		softens, d becomes brown-black	sac	I	yellow, sooty, lower edge of flame is slightly green	hydrochloric acid (HCl) and also a burnt odor	High temp. resistant, 60–67% Cl	PVCC
+	+	+		sac	I/II		Cl, differentiate materials according to Cl-content and softening temp.	Impact resistant to high impact resistant: made elastic with EVAC (or similar)	PVC-HI
+	+			sac	I			made elastic with chlorinated PE	
+	+		m, becomes brown	sac	I/II	yellow, bright, sooty	HCl + paraffin	Chlorinated PE (pure homopolymer)	PEC
	+		similar to PVC	sac	I/II	bright (due to plasticizer)	HCl + plasticizer	stiffens on extracting the plasticizer — Plasticized (properties depending on plasticizer)	PVC-P
	+		becomes clear, doesn't melt, d at red heat	sac	0	doesn't burn, blue-green edge on flame, doesn't char	at red heat stinging odor: HF	F — PFEP, PFA melt at 360 °C, ETFE melts at 270 °C — Polytetrafluoroethylene / PTFE-like molding materials	PTFE, PFEP, PFA, ETFE
	+	+	m, d at red heat	sac	0	like PTFE, sparks	HCl + HF	F, Cl — Polytrifluorochloro-ethylene	CTFE
		+	m, d at high temp.	sac	0/1	hardly flammable	stinging (HF)	F — Polyvinylidene fluoride	PVF_2

Physical Properties (continued)

Standard Abbreviation (ISO 1043/ASTM 1600)	Name	Density unfilled g/cm^3	transparent thin films	transparent, clear	hazy to opaque	usually contains fillers	Gasoline	Toluene	Methylene chloride	Diethylether	Acetone	Ethylacetate	Ethylalcohol	Water
			solid products				s = soluble, sw = swellable, i = insoluble							
4 Polyvinylacetate and Derivatives, Polymethylacrylates														
PVAC	Polyvinyl acetate	1.18	most in dispersion				i	s	s	sw	s	s	s	i
PVAL	Polyvinyl alcohol	1.2–1.3	+				i	i	i	i	i	i	i	s[1]
PVB	Polyvinyl butyral	1.1–1.2	safety glass sheet				i	sw	sw/s	i	sw/s	sw/s	s	i
	Polyacrylates	1.1–1.2	Cop.-dispersion				i/s	s	s	i	s	s	s	i[1]
PMMA	Polymethyl methacrylate	1.18		+			i	s	s	i	s	s	i	i
AMMA	Methyl methacrylate/ acrylonitrile copolymer	1.17		+	yellow		i	i	i	i	i	i	i	i
5 Polymers with Heteroatom Chain Structure (Heteropolymers)														
POM	Polyoxymethylene and similar acetal resins	1.41			+		i	i	i	i	i	i	i	i
PPO	Polyphenylenoxide (modified)	1.06			+		i	s	s	i	i	i	i	i
PC	Polycarbonate	1.20	+	+			i	sw	s	sw	sw	sw	i	i

[1] (PVAL) In absence of acetyl groups also swells in hot water

[1] (Polyacrylates) Polyacrylic acid: s

Pyrolysis and Identification Tests (continued)

leathery or rubbery, soft	flexible, resilient	hard	Sample slowly heated in pyrolysis tube m = melts d = decomposes al = alkaline n = neutral ac = acidic sac = strongly acidic	Reactions of vapors given off	Ignition with small flame 0 = hardly ignitable I = burns in flame, extinguishes in absence of flame II = continues to burn after ignition III = burns vigorously, fulminates		Odor of vapors given off on heating in pyrolysis tube or after ignition and extinction	Characteristic elements (N, Cl, F, S, Si), individual identification tests (Compare Chapter 4 "Testing for Heteroatoms" and Chapter 6 "Specific Identification Tests")	Plastic Materials (arranged in chemical groups) Name	Standard Abbreviation (ISO 1043/ASTM 1600)
						4 Polyvinylacetate and Derivatives, Polymethylacrylates				
+	+		m, brown, vaporizes	ac	II	bright, sooty	acetic acid and additional odor		Polyvinyl acetate	PVAC
+	+		m, d, brown residue	n	I/II	bright	irritating		Polyvinyl alcohol	PVAL
+			m, d, foams	ac	II	blue with yellow edge	rancid butter		Polyvinyl butyral	PVB
+	+		m, d, vaporizes	n	II	bright, slightly sooty	typically sharp		Polyacrylates	
		+	softens, d, swells up and crackles, little residue	n	II	burns with crackling, drips, bright	typically fruity	cast acrylic sheet hardly softens	Polymethyl methacrylate	PMMA
		+	brown, then m, d, black	al	II	sooty, sparks slightly	first sharp, irritating	N	Methyl methacrylate/ acrylonitrile copolymer	AMMA
						5 Polymers with Heteroatom Chain Structure (Heteropolymers)				
		+	m, d, vaporizes	n (ac)	II	blue, almost colorless	formaldehyde		Polyoxymethylene and similar acetal resins	POM
		+	becomes black, m, d, brown vapors	al	II	difficult to ignite, then bright, sooty flame	first slight then phenol odor		Polyphenylene-oxide (modified)	PPO
		+	m, viscous, colorless d, brown	(ac)	I	bright, sooty, bubbly, chars	first slight then phenol odor	Indophenol Test	Polycarbonate	PC

Physical Properties (continued)

Standard Abbreviation (ISO 1043/ASTM 1600)	Name	Density unfilled g/cm³	transparent thin films	transparent, clear	hazy to opaque	usually contains fillers	Gasoline	Toluene	Methylene chloride	Diethylether	Acetone	Ethylacetate	Ethylalcohol	Water
			Usual Appearance (solid products)				Solubility in Cold Solvents (ca. 20 °C) s = soluble, sw = swellable, i = insoluble							
5 Polymers with Heteroatom Chain Structure (Heteropolymers) (continued)														
PET	Polyethylene terephthalate	1.35	+	+	+		i	i	sw	i	i/sw	sw	i	i
PBT	Polybutylene terephthalate	1.41												
PA	Polyamides (crystalline) PA 46	1.14	+		+		i	i	i	i	i	i	i	i
	to PA 12	1.02												
	(amorphous)	1.12		+			i	i	sw	i	sw	i	i	i
PSU	Polysulfone	1.24		(+)	+		i	s	s	i	sw	i/sw	i	i
PI	Polyimides	~1.4	+		+ (yellow)		i	i	i	i	i	i	i	i
CA	**Cellulose Derivatives** Cellulose acetate	1.3	+	+			i	i	sw/s[1]	i	sw/s[1]	i/s[1]	i	i
CAB	Cellulose acetobutyrate	1.2	+	+			i	sw	s	i	s	s	sw	i
CP	Cellulose propionate	1.2		+			i	i	sw	i	s	s	sw	i

[1] depending on degree of acetylation

Pyrolysis and Identification Tests (continued)

Elastic Behavior — leathery or rubbery, soft	flexible, resilient	hard	Sample slowly heated in pyrolysis tube (m = melts, d = decomposes) (al = alkaline, n = neutral, ac = acidic, sac = strongly acidic)	Reactions of vapors given off	Ignition with small flame (0 = hardly ignitable, I = burns in flame, extinguishes in absence of flame, II = continues to burn after ignition, III = burns vigorously, fulminates)	Odor of vapors given off on heating in pyrolysis tube or after ignition and extinction	Characteristic elements (N, Cl, F, S, Si), individual identification tests (Compare Chapter 4 "Testing for Heteroatoms" and Chapter 6 "Specific Identification Tests")	Plastic Materials (arranged in chemical groups) — Name	Standard Abbreviation (ISO 1043/ASTM 1600)
5 Polymers with Heteroatom Chain Structure (Heteropolymers) (continued)									
}	+	+	m, d, dark brown white deposit above	ac	I/II	bright, crackly, drips, sooty	sweetish, irritating	PET melts at 255 °C, PBT melts at 220 °C	Polyethylene terephthalate / Polybutylene terephthalate	PET / PBT
+	+	+	becomes clear, m / d, brown	ac	I/II	difficult to ignite, blueish yellow edge, crackly, drips, fiber forming	characteristic odor similar to burnt horn	N, differentiate by quant. analysis melting ranges: PA 46: 295 °C, PA 66: 225 °C, PA 6: 220 °C, PA 11: 185 °C, PA 12: 180 °C	Polyamides	PA
	+	+	m, bubbly, vapors invisible, brown	sac	II	difficult to ignite, yellow, sooty, chars	first slight amount, finally H_2S		Polysulfone	PSU
		+	doesn't m, brown on strong heating, glows	al	0	glows	on strong heating phenol		Polyimides	PI
								Cellulose Derivatives	
	+	+	m, d, black	ac	II	m, drips yellow green with sparks	acetic acid + burnt paper		Cellulose acetate	CA
		+	m, d, black	ac	II	bright yellow, drops burn as they fall	acetic acid, butyric acid		Cellulose acetobutyrate	CAB
		+	m, d, black	ac	II	same as CAB	propionic acid, burnt paper		Cellulose propionate	CP

Physical Properties (continued)

Standard Abbreviation (ISO 1043/ASTM 1600)	Name	Density unfilled g/cm³	Usual Appearance (solid products) transparent thin films	transparent, clear	hazy to opaque	usually contains fillers	Solubility in Cold Solvents (ca. 20 °C) s = soluble sw = swellable i = insoluble — Gasoline	Toluene	Methylene chloride	Diethylether	Acetone	Ethylacetate	Ethylalcohol	Water
5 Polymers with Heteroatom Chain Structure (Heteropolymers) (continued)														
CN	Cellulose nitrate (celluloid)	1.35–1.4	+	+			i	i	i	sw	s	s	i	i
CMC	(Hydroxy)methylcellulose	>1.29	adhesive raw material				i	i	i	i	i	i	i	s¹
										[1]only cold				
	Cellophane (regenerated cellulose)	1.45	+				i	i	i	i	i	i	i	i¹
											[1]softens			
Vt	Vulcanized fiber	1.2–1.3			+		i	i	i	i	i	i	i	i
6 Phenolic Resins PF = Phenol formaldehyde; including cresol containing resins														
PF	Free of filler: uncured	1.25 to	industrial resins				i	i	i	i	s	i	s	(s)
	molded or cast resins	1.3		(+)			i	i	i	i	i	i	i	i
PF	Mineral filled moldings					+								
PF	Organic filled moldings					+	i	i	i	i	i	i	i	i
PF	Paper based laminates					+								
PF	Cotton based laminates					+								
PF	Asbestos or glass fiber based laminates					+								

Pyrolysis and Identification Tests (continued)

Elastic Behavior — leathery or rubbery, soft	flexible, resilient	hard	Sample slowly heated in pyrolysis tube (m = melts, d = decomposes)	Reactions of vapors given off (al = alkaline, n = neutral, ac = acidic, sac = strongly acidic)	Ignition with small flame (0 = hardly ignitable, I = burns in flame, extinguishes in absence of flame, II = continues to burn after ignition, III = burns vigorously, fulminates)	Odor of vapors given off on heating in pyrolysis tube or after ignition and extinction	Characteristic elements (N, Cl, F, S, Si), individual identification tests (Compare Chapter 4 "Testing for Heteroatoms" and Chapter 6 "Specific Identification Tests")	Plastic Materials (arranged in chemical groups) Name	Standard Abbreviation (ISO 1043/ASTM 1600)
			5 Polymers with Heteroatom Chain Structure (Heteropolymers) (continued)						
+	+		d, violent	sac	III	bright, violent, brown vapors	nitrous oxides (camphor)	N	Cellulose nitrate (celluloid)	CN
			m, chars	n	II	yellow, bright	burnt paper		(Hydroxy)methyl-cellulose	CMC
+			d, chars	n	II	like paper	burnt paper		Cellophane (regenerated cellulose)	
+	+		d, chars	n	I/II	burns slowly	burnt paper		Vulcanized fiber	Vf
			6 Phenolic Resins						
+			m, d / d, cracks	n	I	difficult to ignite, bright, sooty	phenol, formaldehyde		Free of filler: uncured molded or cast resins	PF
	+		d, cracks	n (al)	0/I	bright, sooty	phenol, formaldehyde, possibly ammonia		Mineral filled moldings	PF
	+		d, cracks	n (al)	I/II	chars	as above + burnt paper		Organic filled moldings	PF
	+		d, delamination	n	II	bright, sooty	as above + burnt paper		Paper based laminates	PF
	+								Cotton based laminates	PF
	+		d, cracks	n	0/I	reinforcing structure remains	phenol, formaldehyde		Asbestos or glass fiber based laminates	PF

Physical Properties (continued)

Standard Abbreviation (ISO 1043/ASTM 1600)	Name	Density unfilled g/cm³	transparent thin films	transparent, clear	hazy to opaque	usually contains fillers	Gasoline	Toluene	Methylene chloride	Diethylether	Acetone	Ethylacetate	Ethylalcohol	Water
7 Amino Resins (UF = Urea/Formaldehyde, MF = Melamine/Formaldehyde)														
UF/MF	Uncured		glues				i	i	i	i	i	i	i	s
UF/MF	Organic filled moldings					+								
MF	Mineral filled moldings					+								
MF + PF	Organic filled moldings					+	i	i	i	i	i	i	i	i
MF	Glass fiber fabric based laminates					+								
8 Crosslinked Reaction Resins (UP = unsaturated polyesters, EP = Epoxy resins)														
UP	Unfilled cast resins (flame retarded)	~1.2 (≧1.3)		+			i	i	sw	i	sw	sw	i	i
UP	Moldings, laminates				+	+								
EP	Cast resins (unfilled)	~1.2		+			i	i	sw	i	sw	sw	i	i
EP	Moldings, laminates					+								

Solubility in Cold Solvents (ca. 20 °C)

solid products   s = soluble
sw = swellable
i = insoluble

Pyrolysis and Identification Tests (continued)

Elastic Behavior (leathery or rubbery, soft / flexible, resilient / hard)			Sample slowly heated in pyrolysis tube m = melts d = decomposes al = alkaline n = neutral ac = acidic sac = strongly acidic	Reactions of vapors given off	Ignition with small flame 0 = hardly ignitable I = burns in flame, extinguishes in absence of flame II = continues to burn after ignition III = burns vigorously, fulminates	Odor of vapors given off on heating in pyrolysis tube or after ignition and extinction	Characteristic elements (N, Cl, F, S, Si), individual identification tests (Compare Chapter 4 "Testing for Heteroatoms" and Chapter 6 "Specific Identification Tests")	Plastic Materials (arranged in chemical groups) Name	Standard Abbreviation (ISO 1043/ASTM 1600)
								7 Amino Resins	
+			d, cracks, darkens, swells up	al	0/I	very difficult to ignite, flame slightly yellow, material chars with white edges	ammonia, amines, disgusting fishy odor (esp. with thiourea), formaldehyde	N, possibly S	Uncured	UF/MF
+									Organic filled moldings	UF/MF
									Mineral filled moldings	MF
+									Organic filled moldings	MF + PF
+									Glass fiber fabric based laminates	MF
									8 Crosslinked Reaction Resins	
+			darkens, m cracks, d possibly white deposit above	n (ac)	II (I)	bright, yellow, sooty, softens if no filler, otherwise crackles, chars, filler or glass fiber residue	styrene and sharp additional odor	ignitability depends also on fillers and pigments	Unfilled cast resins (flame retarded)	UP
+									Moldings, laminates	UP
	+		darkens from edge, d, cracks, possibly white deposit above	o or al	II I/II	difficult to ignite, burns with small yellow flame, sooty	depends on curing agent ester-like or amines (similar to PA), later phenol	N with amine curing agents	Cast resins (unfilled)	EP
	+								Moldings, laminates	EP
			(uncross-linked)	(ac)						

Physical Properties (continued)

Standard Abbreviation (ISO 1043/ASTM 1600)	Name	Density unfilled g/cm³	transparent thin films	transparent, clear	hazy to opaque	usually contains fillers	Gasoline	Toluene	Methylene chloride	Diethylether	Acetone	Ethylacetate	Ethylalcohol	Water
9 Polyurethanes														
PUR	Crosslinked	1.26			+		i	i	sw	i	sw	sw	i	i
	Linear, rubberlike	1.17–1.22	+		+		i	sw	sw	i	sw	sw	i	i
10 Silicones														
SI	Mainly silicone rubber	1.25				+	sw	sw	sw	i	i	i	i	i
	Prepolymers (Silicones)						s						s	

Solubility in Cold Solvents (ca. 20 °C)
s = soluble
sw = swellable
i = insoluble

solid products

8.2 Chemicals

The chemicals needed to carry out the tests described earlier are listed in this chapter. They are available from commercial suppliers. We recommend that the most important acids, bases, and solvents be ordered in at least 0.5–1-liter quantities. Dilute solutions can be prepared in the laboratory. As for indicator reagents, it is generally sufficient to order 1–5 g. For storing chemicals, use only unambiguously identified bottles unless the reagents are supplied in labeled plastic containers.

It must be pointed out again that many organic solvents are flammable and should therefore be stored only in small amounts. The use of concentrated acids and bases also requires special safety measures, since they can cause injuries to skin and eyes.

Pyrolysis and Identification Tests (continued)

Elastic Behavior			Sample slowly heated in pyrolysis tube m = melts d=decomposes al = alkaline n = neutral ac = acidic sac = strongly acidic	Reactions of vapors given off	Ignition with small flame 0 = hardly ignitable I = burns in flame, extinguishes in absence of flame II = continues to burn after ignition III = burns vigorously, fulminates		Odor of vapors given off on heating in pyrolysis tube or after ignition and extinction	Characteristic elements (N, Cl, F, S, Si), individual identification tests (Compare Chapter 4 "Testing for Heteroatoms" and Chapter 6 "Specific Identification Tests")	Plastic Materials (arranged in chemical groups) Name	Standard Abbreviation (ISO 1043/ASTM 1600)
leathery or rubbery, soft	flexible, resilient	hard								
									9 Polyurethanes	
+	+	+	m on strong heating, then d	al		II	difficult to ignite, yellow, bright foams, drips	typically unpleasant stinging (iso-cyanate)	N	Crosslinked	
+				ac						Linear, rubberlike	
										10 Silicones	
+			d only on strong heating, white powder	n		0	glows in the flame	white smoke, finely divided white SiO_2 residue	Si	Mainly silicone rubber Prepolymers (Silicones)	SI

All solvents and chemicals named here are available in several degrees of purity, for example, technical, pure, chemically pure, for analysis (p.a.), etc. As far as possible, use only analytically pure reagents. Solvents that have turned yellow or dark on storage must be distilled before use.

Acids and Bases

Table 25 provides directions for the preparation of the required dilute solutions from the commercially available concentrated solutions. If it is not otherwise indicated, the dilute solutions used in this book refer to approximately 2 normal ($2N$) solutions. In diluting a concentrated acid or base, always add the acid or base to the required amount of distilled or deionized

Table 25 Concentrations and Densities of Commercially Available Acids and Bases

Acid or Base	Content in		
	Weight %	Mol/liter	Normality
Concentrated sulfuric acid ($d = 1.84\,\mathrm{g/cm^3}$)	96		37
Dilute sulfuric acid	9	1	2
Fuming nitric acid	86		
Concentrated nitric acid ($d = 1.40\,\mathrm{g/cm^3}$)	65	10	10
Dilute nitric acid	12	2	2
Fuming hydrochloric acid ($d = 1.19\,\mathrm{g/cm^3}$)	38	12.5	12.5
Concentrated hydrochloric acid ($d = 1.16\,\mathrm{g/cm^3}$)	32	10	10
Dilute hydrochloric acid	7	2	2
Glacial acetic acid (water-free)	100		17
Dilute acetic acid	12	2	2
Dilute sodium hydroxide	7.5	2	2
Concentrated ammonia	25	13	6.5
Dilute ammonia	3.5	2	2

water, never the other way around, since the resulting heat can lead to spattering. (*Always wear safety glasses!*)

Dilute solutions are prepared in the laboratory according to the following directions:

Dilute sulfuric acid: 5 ml concentrated acid
  ($d = 1.84\,\mathrm{g/cm^3}$) in 90 ml water
Dilute nitric acid: 13 ml concentrated acid
  ($d = 1.40\,\mathrm{g/cm^3}$) in 80 ml water
Dilute hydrochloric acid: 19 ml concentrated acid
  ($d = 1.16\,\mathrm{g/cm^3}$) in 80 ml water
Dilute acetic acid: 12 ml glacial acetic acid
  in 88 ml water
Dilute ammonia: 17 ml concentrated ammonia
  ($d = 0.882\,\mathrm{g/cm^3}$) in 90 ml water
Dilute sodium hydroxide: Dissolve 8 g hydroxide
  in 100 ml water

In addition to the acids and bases mentioned in Table 25 you will often need:

Acetic anhydride
Formic acid
3% Hydrogen peroxide

For the preparation of all aqueous solutions, always use distilled or deionized water, never tap water. Polyethylene squeeze bottles (250 ml capacity) are usually very practical, since they are very suitable for the storage of distilled water and methanol.

Inorganic Chemicals

Zinc chloride, anhydrous   } for density
Magnesium chloride, anhydrous   } determination
Iron(II) sulfate
Iron(III) chloride (1.5 N solution in water)
Sodium nitroprussate
Lead acetate as a 2 N solution
    (26.7 g to 100 g of water)
Silver nitrate as a 2% solution (store in the dark)
Calcium chloride
Sodium hydroxide
Ammonium molybdate
Ammonium sulfate
Sodium carbonate (anhydrous)
Sodium rhodizonate
Sodium hydroxide
Sodium peroxide
Sodium nitrite
Sodium acetate
Sodium thiosulfate
Sodium hypochlorite or chloride of lime solution
Barium chloride

Potassium hydroxide (2.8 g to 100 g water)
Potassium iodide
Mercury(II) oxide
Nickel sulfate
Copper(II) acetate
Lanthanum nitrate
Borax
Barium hydroxide solution,
  approximately 0.2 N (1.7 g to 100 g water)
Sodium or potassium under petroleum
  or some other inert liquid
Iodine
0.1 N iodine-potassium iodide solution: dissolve 16.7 g potassium
  iodide in 200 ml water, then dissolve 12.7 g of iodine in this
  solution and dilute the mixture with water to 1000 ml)
Wijs solution or iodine monochloride (see Section 6.2.18)
Zinc dust

Organic Solvents

Not all the organic solvents listed in Table 6 are necessary. The
supply can be limited to the following:

Toluene
p-Xylene
Nitrobenzene
n-Hexane or petroleum ether
Cyclohexanone
Tetrahydrofuran
Dioxane
Diethyl ether
Formamide
Dimethylformamide
Dimethylsulfoxide
Chloroform

Carbon tetrachloride
Methanol
Ethanol
Ethylene glycol
Acetone
Ethyl acetate
m-Cresol
Benzyl alcohol
Benzylamine
Pyridine

Organic Reagents

Benzidine
2,2'-Bipyridyl
Diphenylamine
Chromotropic acid
Thymol
Morpholine
Hydroquinone
o-Nitrobenzaldehyde
p-Dimethylaminobenzaldehyde
2,6-Dibromoquinone-4-chloroimide
α-Naphthol
4-Nitrobenzodiazoniumfluoroborate
   (Nitrazol CF-extra)
Phenylhydrazine
Urease

Miscellaneous

Litmus paper (red and blue)
Congo red paper
pH paper, as a universal paper for many experiments

Lead acetate paper (keep in a closed, dark bottle)
Cotton wool
Glass wool
Fine sand
Silver coin (for sulfur identification)
Copper wire
Active charcoal

Important reminder: Please refer to the safety instructions in your laboratory for the safe handling and storage of all solvents and chemicals. Always wear safety glasses and protective clothing when working in the laboratory; avoid skin contact with chemicals and solvents (wear protective gloves!).

8.3 Laboratory Aids and Equipment

The tests described in this book generally do not require any apparatus or equipment other than normal laboratory equipment. The following list contains what might have to be ordered if a laboratory is not available.

For heating, use a hot plate or a heating mantle whenever possible. Open flames, a Bunsen burner, or, if there is no gas connection, an alcohol or Sterno burner should be used only when it is necessary to heat the samples in test tubes or pyrolysis tubes. For flame tests, a candle is sufficient.

Basic Equipment

Safety glasses, protective clothing, protective gloves
Test tubes, small, approximately 7 mm in diameter; medium, approximately 15 mm diameter
Corks or rubber stoppers to fit the test tubes
Beakers, 50, 100, 250, 1000 ml

Glass funnels,
    approximately 4 and 7 cm in diameter
Watch glasses
Combustion tubes, approximately 8×70 mm
Glass rods
Graduated cylinders, 10 ml, 100 ml, 500 ml
Pipets, 1 ml, 10 ml
Porcelain mortar with pestle,
    approximately 10 cm in diameter
Porcelain plate
Porcelain cups, approximately 5 cm in diameter
Porcelain crucible,
    approximately 3–3.5 cm in diameter
Platinum or nickel crucible,
    approximately 3 cm in diameter
Aerometer for density measurement
    in the range 0.8–2.2 g/cm^2
A small balance
    (if nothing better is available, a letter scale is sufficient)
Test tube rack
Test tube tongs
Crucible tongs
Tweezers
Spatula
Knife
Filter paper in sheets and round filter paper for the funnels
Oil bath of metal (best, a silicon oil)

Optional Equipment

Mill for grinding plastic samples
Heating mantles and a stand with clamps
Distillation flask, reflux condenser (see Figure 9)
Soxhlet apparatus with tubes for extraction (see Figure 2)
Hot stage (see Figure 6) or melting point microscope.

8.4 Guide for Further Reading

Bark, L. S., Allen, N. S. (Eds.):
Analysis of Polymer Systems.
Applied Science Publishers Ltd., London, 1982.

Compton, T. R.:
Chemical Analysis of Additives in Plastics, 2nd Ed.
Pergamon, Oxford, New York, 1977.

Ezrin, M.:
Plastics Failure Guide. Cause and Prevention.
Hanser Publishers, Munich, Cincinnati, 1996.

Garton, A.:
Infrared Spectroscopy of Polymer Blends, Composites and Surfaces.
Hanser Publishers, Munich, Cincinnati, 1992.

Haslam, J., Willis, H. A., Squirrel, D. C. M.:
Identification and Analysis of Plastics, 2nd Ed.
Butterworth, London, 1972; Paperback Reprint Edition, Heyden & Son, London and Philadelphia, 1980.

Hummel, D. O., Scholl, F.:
Atlas of Polymer and Plastics Analysis, 2nd Revised Ed. (3 vols.):
Vol. 1: Polymers, Structures and Spectra. Vol. 2a: Plastics, Fibers, Rubbers, Resins, Starting and Auxiliary Materials, Degradation Products. Vol. 2b: Spectra. Vol. 3: Additives and Processing Aids.
Carl Hanser Verlag, Munich, Vienna/VCH (Wiley-VCH) Weinheim, New York, 1978, 1981, 1985, 1986.

Kämpf, G.:
Characterization of Plastics by Physical Methods. Experimental Techniques and Practical Applications.
Hanser, Munich, Vienna, 1986.

Kline, G. M. (Ed.):
High Polymers, Vol. XIII, Analytical Chemistry of Polymers.
(3 Parts).
Interscience, New York, 1959, 1962. (John Wiley & Sons).

Krause, A., Lange, A., Ezrin, M.:
Plastics Analysis Guide. Chemical and Instrumental Methods.
Hanser, Munich, Vienna, 1983.

Mitchell, J. Jr. (Ed.):
Applied Polymer Analysis and Characterization.
Hanser, Munich, Vienna, 1987.

Schröder, E.:
Guide to Polymer Characterization.
Hanser, Munich, Vienna, 1988.

Wake, W. C.:
Analysis of Rubbers and Rubber-like Polymers. 2nd Ed.
Maclaren, London 1969.

8.5 Polymer Acronyms

ABR	Acrylate-butadiene rubber
ABS	Acrylonitrile-butadiene-styrene rubber
ACM	Acrylate rubber
AES	Acrylonitrile-ethylene-propylene-styrene quaterpolymer
AMMA	Acrylonitrile-methyl methacrylate copolymer
ANM	Acrylonitrile-acrylate rubber
APP	Atactic polypropylene
ASA	Acrylonitrile-styrene-acrylate terpolymer

BIIR	Brominated isobutene-isoprene (butyl) rubber
BR	Cis-1,4-butadiene rubber (cis-1,4-polybutadiene)
BS	Butadiene-styrene copolymer (see also SB)
CA	Cellulose acetate
CAB	Cellulose acetate-butyrate
CAP	Cellulose acetate-propionate
CF	Cresol-formaldehyde resin
CHC	Epichlorohydrin-ethylene oxide rubber
CHR	Epichlorohydrin rubber (see also CO)
CMC	Carboxymethyl cellulose
CN	Cellulose nitrate (see also NC)
CNR	Carboxynitroso rubber; (tetrafluoroethylene-trifluoronitrosomethane-unsat.monomer terpolymer)
CO	Poly[(chloromethyl)oxirane]; epichlorohydrin rubber (see also CHR)
CP	Cellulose propionate
CPE	Chlorinated polyethylene
CR	Chloroprene rubber
CS	Casein
CSM	Chlorosulfonated polyethylene
CTA	Cellulose triacetate
CTFE	Poly(chlorotrifluoroethylene); (see also PCTFE)
EAA	Ethylene-acrylic acid copolymer
EAM	Ethylene-vinyl acetate copolymer
EC	Ethyl cellulose
ECB	Ethylene copolymer blends with bitumen
ECTFE	Ethylene-chlorotrifluoroethylene copolymer
EEA	Ethylene-ethyl acrylate copolymer
EMA	Ethylene-methacrylic acid copolymer or ethylene-maleic anhydride copolymer
EP	Epoxy resin
E/P	Ethylene-propylene copolymer (see also EPM, EPR)

EPDM	Ethylene-propylene-nonconjugated diene terpolymer (see also EPT)
EPE	Epoxy resin ester
EPM	Ethylene-propylene rubber (see also E/P, EPR)
EPR	Ethylene-propylene rubber (see also E/P, EPM)
EPS	Expanded polystyrene; polystyrene foam (see also XPS)
EPT	Ethylene-propylene-diene terpolymer (see also EPDM)
ETFE	Ethylene-tetrafluoroethylene copolymer
EVA, E/VAC	Ethylene-vinyl acetate copolymer
EVE	Ethylene-vinyl ether copolymer
FE	Fluorine-containing elastomer
FEP	Tetrafluoroethylene-hexafluoropropylene rubber; see PFEP
FF	Furan-formaldehyde resins
FPM	Vinylidene fluoride-hexafluoropropylene rubber
FSI	Fluorinated silicone rubber
GR-I	Butyl rubber (former US acronym) (see also IIR, PIBI)
GR-N	Nitrile rubber (former US acronym) (see also NBR)
GR-S	Styrene-butadiene rubber (former US acronym; see PBS, SBR)
HDPE	High density polyethylene
HEC	Hydroxyethylcellulose
HIPS	High impact polystyrene
HMWPE	High molecular weight polyethylene
IIR	Isobutene-isoprene rubber; butyl rubber (see also GR-I, PIBI)
IPN	Interpenetrating polymer network
IR	Synthetic cis-1,4-polyisoprene

LDPE	Low density polyethylene
LLDPE	Linear low density polyethylene
MABS	Methyl methacrylate-acrylonitrile-butadiene-styrene
MBS	Methyl methacrylate-butadiene-styrene terpolymer
MC	Methyl cellulose
MDPE	Medium density polyethylene (ca. $0.93-0.94\,g/cm^3$)
MF	Melamine-formaldehyde resin
MPF	Melamine-phenol-formaldehyde resin
NBR	Acrylonitrile-butadiene rubber; nitrile rubber; GR-I
NC	Nitrocellulose; cellulose nitrate (see also CN)
NCR	Acrylonitrile-chloroprene rubber
NIR	Acrylonitrile-isoprene rubber
NR	Natural rubber (cis-1,4-polyisoprene)
OER	Oil extended rubber
PA	Polyamide (e.g. PA 6,6 = polyamide, 6,6 = Nylon 6,6 in US literature)
PAA	Poly(acrylic acid)
PAI	Polyamide-imide
PAMS	Poly(alpha-methylstyrene)
PAN	Polyacrylonitrile (fiber)
PARA	Poly(acrylamide)
PB	Poly(1-butene)
PBI	Poly(benzimidazole)
PBMA	Poly(n-butyl methacrylate)
PBR	Butadiene-vinyl pyridine copolymer
PBS	Butadiene-styrene copolymer (see also GR-S, SBR)
PBT, PBTP	Poly(butylene terephthalate)
PC, PCO	Polycarbonate
PCD	Poly(carbodiimide)
PCTFE	Poly(chlorotrifluoroethylene)
PDAP	Poly(diallyl phthalate)
PDMS	Poly(dimethylsiloxane)

PE	Polyethylene
PEA	Poly(ethyl acrylate)
PEC	Chlorinated polyethylene (see also CPE)
PEEK	Poly(arylether ketone)
PEI	Poly(ether imide)
PEO, PEOX	Poly(ethylene oxide)
PEP	Ethylene-propylene polymer (see also E/P, EPR)
PEPA	Polyether-polyamide block copolymer
PES	Polyethersulfone
PET, PETP	Poly(ethylene terephthalate)
PF	Phenol-formaldehyde resin
PFA	Perfluoroalkoxy resins
PFEP	Tetrafluoroethylene-hexafluoropropylene copolymer; FEP
PI	Polyimide
PIB	Polyisobutylene
PIBI	Isobutene-isoprene copolymer; butyl rubber; GR-I, IIR
PIBO	Poly(isobutylene oxide)
PIP	Synthetic cis-1,4-polyisoprene; (also CPI, IR)
PIR	Polyisocyanurate
PMA	Poly(methyl acrylate)
PMI	Polymethacrylimide
PMMA	Poly(methyl methacrylate)
PMMI	Polypyromellitimide
PMP	Poly(4-methyl-1-pentene)
PO	Poly(propylene oxide); or polyolefins; or phenoxy resins
POM	Polyoxymethylene, polyformaldehyde
POP	Poly(phenylene oxide) (also PPO/PPE)
POR	Propylene oxide rubber
PP	Polypropylene
PPC	Chlorinated polypropylene
PPE	Poly(phenylene ether)
PPMS	Poly(para-methylstyrene)

PPO	Poly(phenylene oxide) (also PPO/PPE)
PPOX	Poly(propylene oxide)
PPS	Poly(phenylene sulfide)
PPSU	Poly(phenylene sulfone)
PPT	Poly(propylene terephthalate)
PS	Polystyrene
PSB	Styrene-butadiene rubber (see GR-S, SBR)
PSF, PSO	Polysulfone
PSU	Poly(phenylene sulfone)
PTFE	Poly(tetrafluoroethylene)
P3FE	Poly(trifluoroethylene)
PTMT	Poly(tetramethylene terephthalate) = poly(butylene terephthalate) (see also PBT, PBTP)
PUR	Polyurethane
PVA, PVAC	Poly(vinyl acetate)
PVAL	Poly(vinyl alcohol) (also PVOH)
PVB	Poly(vinyl butyral)
PVC	Poly(vinyl chloride)
PVCA	Vinyl chloride-vinyl acetate copolymer (also PVCAC)
PVCC	Chlorinated poly(vinyl chloride)
PVDC	Poly(vinylidene chloride)
PVDF	Poly(vinylidene fluoride)
PVF	Poly(vinyl fluoride)
PVFM	Poly(vinyl formal) (also PVFO)
PVI	Poly(vinyl isobutyl ether)
PVK	Poly(N-vinylcarbazole)
PVP	Poly(N-vinylpyrrolidone)
RF	Resorcinol-formaldehyde resin
SAN	Styrene-acrylonitrile copolymer
SB	Styrene-butadiene copolymer
SBR	Styrene-butadiene rubber (see also GR-S)
SCR	Styrene-chloroprene rubber
S-EPDM	Sulfonated ethylene-propylene-diene terpolymers

SHIPS	Super-high impact polystyrene
SI	Silicone resins; poly(dimethylsiloxane)
SIR	Styrene-isoprene rubber
SMA	Styrene-maleic anhydride copolymer
SMS	Styrene-alpha-methylstyrene copolymer
TPE	Thermoplastic elastomer
TPR	1,5-trans-Poly(pentenamer)
TPU	Thermoplastic polyurethane
TPX	Poly(methyl pentene)
UF	Urea-formaldehyde resins
UHMW-PE	Ultrahigh molecular weight poly(ethylene) (also UHMPE) (molecular mass over 3.1×10^6 g/mol)
UP	Unsaturated polyester
VC/E	Vinyl chloride-ethylene copolymer
VC/E/VA	Vinyl chloride-ethylene-vinyl acetate copolymer
VC/MA	Vinyl chloride-methyl acrylate copolymer
VC/MMA	Vinyl chloride-methyl methacrylate copolymer
VC/OA	Vinyl chloride-octyl acrylate
VC/VAC	Vinyl chloride-vinyl acetate copolymer
VC/VDC	Vinyl chloride-vinylidene chloride
VF	Vulcan fiber
XLPE	Crosslinked polyethylene
XPS	Expandable or expanded polystyrene; (see also EPS)

Index

PIT is the abbreviation for the Plastics Identification Table (Table 24). The numbers in brackets refer to the first column.